Dr. Moeness G. Amin
Center for Advanced Communications
Villanova University E-2

Ultrasonic Methods of Non-destructive Testing

NON-DESTRUCTIVE EVALUATION SERIES

Non-destructive evaluation now has a central place in modern technology both as a means of evaluating materials and products as they are manufactured and for confirmation of fitness for purpose while they are in use.

This series provides in-depth coverage of the wide range of techniques that are now available for the non-destructive evaluation of materials. Each volume will contain material that is relevant to final year undergraduates in Engineering, Materials Science and Physics in addition to post graduate students, experienced research workers, and practising engineers. In some cases they will be written with taught courses in mind, while other texts will be for the qualified engineer or scientist who wishes to become familiar with a new topic at research level.

Series editors

Professor S. Palmer	*Professor W. Lord*
Department of Physics	*Department of Electrical and Computer Engineering*
University of Warwick	*Iowa State University*
Coventry	*Iowa*
UK	*USA*

Titles available

Numerical Modeling for Electromagnetic Non-Destructive Evaluation
Nathan Ida

Industrial Radiology
R. Halmshaw

Ultrasonic Methods of Non-destructive Testing
Blitz and Simpson

Ultrasonic Methods of Non-destructive Testing

Jack Blitz

Associate Senior Lecturer
Department of Physics
Brunel University

and

Geoff Simpson

Consultant and Lecturer in NDT

CHAPMAN & HALL

London · Glasgow · Weinheim · New York · Tokyo · Melbourne · Madras

Published by Champman & Hall, 2–6 Boundary Row, London SE1 8HN, UK

Chapman & Hall, 2–6 Boundary Row, London SE1 8HN, UK

Blackie Academic & Professional, Wester Cleddens Road, Bishopbriggs, Glasgow G64 2NZ, UK

Chapman & Hall GmbH, Pappelallee 3, 69469 Weinheim, Germany

Chapman & Hall USA, 115 Fifth Avenue, New York, NY 10003, USA

Chapman & Hall Japan, ITP-Japan, Kyowa Building, 3F, 2-2-1 Hirakawacho, Chiyoda-ku, Tokyo 102, Japan

Chapman & Hall Australia, 102 Dodds Street, South Melbourne, Victoria 3205, Australia

Chapman & Hall India, R. Seshadri, 32 Second Main Road, CIT East, Madras 600 035, India

First edition 1996

Typeset in 10/12 Palatino by Best-set Typesetter Ltd., Hong Kong. Printed in Great Britain at the University Press, Cambridge.

ISBN 0 412 60470 1

A catalogue record for this book is available from the British Library

Library of Congress Catalog Card Number: 95-70019

∞ Printed on permanent acid-free text paper, manufactured in accordance with ANSI/NISO Z39.48-1992 and ANSI/NISO Z39.48–1984 (Permanence of Paper).

Contents

Preface

This book is intended to provide an account of the theory and practice of **everyday** ultrasonic non-destructive testing although references are made to relevant latest developments, some of which are still in the research stage. It is aimed to cater for a wide range of readers extending from those embarking on associate degrees or national certificate courses to undergraduate and postgraduate students, researchers and operators, engineers and management involved with non-destructive testing.

In the preparation of this work, the authors are indebted to Professor D.D. Imrie, Dean of the Faculty of Science, at Brunel University and Mr C.D. Wells of Messrs. Wells–Krautkramer for the use of their facilities and to Mr M. Berke of Krautkrämer GmbH and Co., Commander G.M. Selous OBE and his colleagues at CNS Electronics Ltd, Dr E.L. Short of the Department of Chemistry at Brunel University and Dr M.G. Silk of the National Nondestructive Testing Centre at Harwell for their kind help.

Jack Blitz, Department of Physics, Brunel University, UK.
Geoff Simpson, Letchworth, UK.

CHAPTER 1

Introduction

GENERAL CONSIDERATIONS

Ultrasonics is the name given to the study and application of **ultrasound**, which is sound of a pitch too high to be detected by the human ear, i.e. of frequencies greater than about 18 kHz. Ultrasonic waves have a wide variety of applications over an extended range of intensity, including cutting, cleaning and the destruction of tissue at the upper extremity and **non-destructive testing (NDT)** at the lower end. A non-destructive test is one in which there is no impairment of the properties and performance in future use of the object under examination. With ultrasonic non-destructive testing, which is effectively a mechanical method, periodic mechanical stresses are applied to the object. It is essential that there have been no changes in dimensions and structure of the object when the test is completed. This can only be achieved when the maximum applied stresses do not exceed the elastic limit below which Hooke's law is obeyed, so that the resultant strain is proportional to the applied stress. Hence it is necessary that the ultrasonic intensity is sufficiently low for the elastic limit not to be exceeded.

Ultrasonic testing consists effectively of the propagation of low amplitude waves through a material to measure either or both the time of travel and any change of intensity for a given distance. Applications include distance gauging, flaw detection and measuring parameters (such as elastic moduli and grain size) which are related to the material structure. Reasons for using ultrasonic as opposed to audible frequencies include the following.

- Wavelengths decrease inversely with frequency. This is advantageous with the testing of smaller samples having dimensions of the same order of magnitude as the wavelength used. The use of short wavelengths enables the employment of shorter pulses, thus providing higher degrees of resolution for defect detection. Furthermore, the degree of beam spread decreases with rise in frequency and, hence, an increase in directivity; this is of great importance for locating defects.
- Attenuation generally increases with frequency with the result that the degree of attenuation, which is often related to material structure,

can be more easily detected and measured. In some instances, such as the testing of highly attenuating materials such as concrete with the resonance and damping capacity method, satisfactory results can be obtained with the use of higher audible frequencies, e.g. 10 kHz.

The application of ultrasound to non-destructive testing was first made possible by the discovery of the piezoelectric effect by the brothers Pierre and Jacques Curie in 1880. However, it was not until World War II when the necessary technology became available, after the discovery of radar, that any effective progress was possible. It was then that Firestone in the USA and Sproule in the UK, working independently of one another, developed the pulse–echo ultrasonic flaw detector. Since then, ultrasonic testing has progressed at a remarkable rate as a result of rapid advancements in electronics and in computer technology. It is now more frequently used than radiology for non-destructive testing.

The ultrasonic method of non-destructive testing is currently the most widely used of the testing methods, disregarding the obvious ones of looking, feeling, measuring and weighing. These 'obvious' tests should, where possible, be conducted initially in all cases. It is a sad fact that much time and money have been wasted by conducting tests which reveal defects that could have been observed by the naked eye. The other principal methods are:

- radiology, e.g. the uses of X- and γ-rays and of neutron beams (for example, Halmshaw 1982);
- electrical and magnetic methods, e.g. magnetic particle inspection and other magnetic flux leakage methods, eddy current testing, potential drop and AC field methods and microwave testing (for example, Blitz 1991); and
- liquid penetrant testing (for example, McMaster 1982).

Of the commonly-used methods, only ultrasonic and radiological techniques can detect internal flaws with a very high degree of certainty although, in limited cases, detection may be possible in dielectric materials with microwaves and in ferromagnetic materials with magnetic flux leakage methods using highly sensitive Hall element detectors. The use of the other methods is mainly restricted to detecting surface and sub-surface defects for which ultrasonic testing is often successful, but it should be borne in mind that these other methods, especially the magnetic particle and dye penetrant methods can be performed more speedily. However, the latter techniques cannot provide reliable, if any, indications of crack depth which can be accurately measured with ultrasound, eddy currents and AC potential drop and field measurements. For metals, the use of eddy currents is preferred for evaluating shallow surface cracks in metals and, although the AC measurements may be used for deeper crack detection and measurements, the ultrasonic technique has the advantage of accurately locating crack tips and thus providing indications of the angles of cracks.

The main advantages of ultrasonic testing are:

- testing can be carried out from a single surface;
- a high degree of penetration is possible in many commonly-used materials, which is in contrast with the lower degree of penetration encountered with radiological testing of metals;
- accuracy in locating and measuring defects;
- the ability to detect and size very small defects; and
- compatibility with automatic scanning devices and with microprocessors and computers.

The principal drawbacks are these:

- Operators must be properly trained, highly experienced and possess a high degree of reliability and integrity. However, with 100% automated testing, these requirements may be somewhat relaxed provided that the functions of the automated system are thoroughly understood.
- With manual operation over a large surface area only a small part of a surface can be scanned at a time, although this can be improved upon, where feasible, with the use of transducer arrays.
- A high degree of coupling between the transducer and the surface to be scanned is required, though much progress is now being made with the development of non-contact transducers.

This book is intended to provide an account of the theory and practice of everyday ultrasonic testing, although references are made to relevant latest developments, some of which are still in the research stage. It is aimed to cater for a wide range of reader extending from those embarking on associate degree or national certificate and BTEC courses to postgraduate researchers and, of course, qualified engineers and NDT personnel. To allow for this, the mathematics are kept at as low a level as possible in the main parts of the text and derivations of the relatively more complex equations are located in the Appendices.

Following normal practice, SI units are used in this book and readers who are not too familiar with them are referred to Appendix H.

REFERENCES

Blitz, J. (1991) *Electrical and Magnetic Methods of Nondestructive Testing*. Adam Hilger, Bristol.

Halmshaw, R. (1995) *Industrial Radiology – Theory and Practice*, 2nd edn Chapman & Hall, London.

McMaster, R.C. (ed) (1982) *Nondestructive Testing Handbook,* 2nd edn, Vol 2 (Liquid Penetrants). American Society for Nondestructive Testing, Columbus, OH.

CHAPTER 2

The propagation of low amplitude ultrasound

2.1 INTRODUCTION

Acoustic waves are radiated when a vibrating body, known as the **source**, applies periodic stresses to a material so as to give rise to periodic strains which are propagated as waves. With ultrasonic non-destructive testing it is important that the magnitudes of these stresses are sufficiently low for the material to remain in its elastic state: the ratio of the stress to the strain remains constant for fixed physical conditions, in accordance with Hooke's law. This is necessary so that:

- the propagation characteristics (speed and attenuation coefficient) of the waves stay constant; and
- the object being tested is not subjected to permanent deformation so as to render the test no longer non-destructive.

Ultrasonic non-destructive testing involves, often indirectly, one or both of the following measurements of the object being examined:

- the time of travel of the waves for a given distance; and
- the amount of attenuation of the waves.

The time of travel depends on the dimensions of the object and on the speed of ultrasound through it; the measured attenuation is related to the attenuation coefficient for the material of the object (section 3.3.1). These quantities are dependent on the nature and structure of the material and also the temperature, pressure and ultrasonic frequency. They may also be dependent on the acoustic intensity, if the object is subjected to elastic hysteresis by the stresses transmitted by the waves (section 3.3.2), although the object may revert to its original state when the ultrasound is removed. One of the advantages of using high frequency sound waves is an increase in directivity (section 2.7), i.e. the beam can be made parallel for relatively long distances when plane-surfaced sources are used to produce plane waves. To maintain a constant frequency free from unwanted harmonics, the source should vibrate with simple harmonic motion (section 2.2).

When a source vibrates in the direction of the required wave motion, perpendicular to its surface, longitudinal waves which are characterized by alternating regions of compression and dilatation, i.e. **compression waves**, are propagated. With a source vibrating at right-angles to the wave direction, the medium of propagation suffers alternating shear stresses for which the waves are transverse and are called **shear waves**. In practice, except for very short distances in highly viscous liquids, shear waves can be produced only in solids. It should be pointed out here that not all transverse waves are shear waves; e.g. the flexural vibrations in cantilevers are mainly caused by alternating tensile and compressive stresses.

2.2 FREE AND FORCED VIBRATIONS

2.2.1 General considerations

Acoustic waves in a body arise from a mechanical disturbance, either periodic or aperiodic, usually from a vibrating source in contact with it (but sometimes from a non-contacting source such as a laser transmitter or an electromagnetic acoustic transducer (section 4.5.1)). The contacting vibrating source may take one of a number of different forms but, in virtually all instances, its operation can be studied by considering the motion of a mass, M, across a horizontal surface. This mass is connected by means of a horizontal light helical spring of mass, m, to a fixed support (Fig. 2.1). The spring has a **stiffness**, S, (i.e. applied force per unit displacement) which remains constant at all times provided that the elastic limit for its material is not exceeded. For reasons which soon become apparent, the **compliance**, $C_m = 1/S$ (i.e. the displacement per unit applied force) is considered here. The frictional force per unit velocity (i.e. the **mechanical resistance**) R_m between the mass and the surface is assumed to be constant for the relatively low values of velocity of the mass, consistent with small strains experienced by the spring.

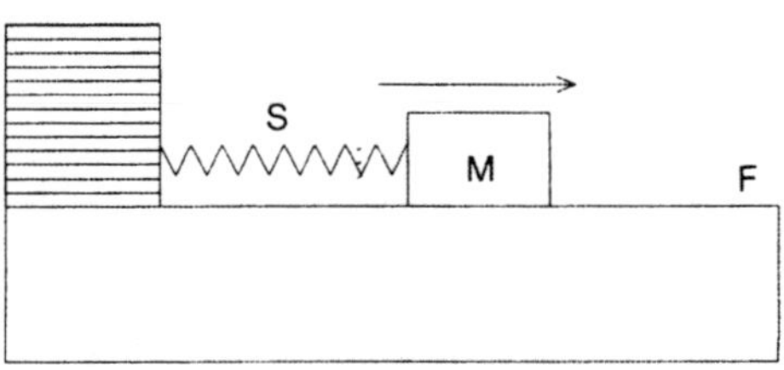

Fig. 2.1 Mass M on horizontal plane surface F and connected to fixed support P by horizontal spring S. Arrow indicates positive direction of y.

2.2.2 Free damped vibrations

Let the mass M be displaced along the axis of the spring by an amount y_0, so as to apply a tensile stress, and then released. Provided that the mechanical resistance is sufficiently small for the parameter $R_m/2M$ to be less than $(1/C_mM)^{1/2}$, the mass vibrates freely with damped harmonic motion and the relationship between the **displacement**, y, from the zero position at a time, t, is shown in Appendix A to be given by:

$$y = y_0(\omega_0/\omega)\exp(-\alpha' t)\cos(\omega t - \phi), \tag{2.1}$$

where

$$\alpha' = -R_m/2M$$

and

$$\omega_0 = (1/C_mM)^{1/2}.$$

y_0 is defined as the **displacement amplitude**, and the expression $\omega t - \phi$ is called the **phase angle**. The time taken for a phase change of 360°, 2π, to take place is known as the **time period**, T, of the vibrations. The reciprocal of T is the **natural frequency**, f. When T is expressed in seconds, the unit of f is the hertz (Hz). ω is called the **angular frequency** and has units s^{-1}. Thus:

$$T = 1/f = 2\pi/\omega. \tag{2.2}$$

The angle ϕ can be eliminated by putting $(\omega t - \phi)$ equal to zero when $t = 0$. Its use is only significant when comparing two or more vibrating systems so as to indicate the **phase difference**. α' is defined here as the **attenuation per unit time**.

Very short pulses of only two or three cycles in duration, as employed with ultrasonic flaw detection, can be generated by the use of a heavily-damped source (section 4.4) for which the value of α' is relatively high. However, if the vibrations are sustained for more than a few cycles, the value of α' may be sufficiently low for ω/ω_0 to approximate to unity. Equation 2.1 can then be written as:

$$y = y_0\exp(-\alpha' t)\cos\omega t, \tag{2.1a}$$

and, for negligible damping, α' tends to zero so that:

$$y = y_0\cos\omega_0 t. \tag{2.1b}$$

$f_0 = \omega_0/2\pi$ is then equal to the **resonant frequency** (section 2.2.4). The variation of the displacement y with time t is shown in Fig. 2.2(a) for zero damping and 2.2(b) for finite damping. The exponential envelopes are seen to touch the curves slightly to the right of the maxima and minima although it should be noted that the approximate equation

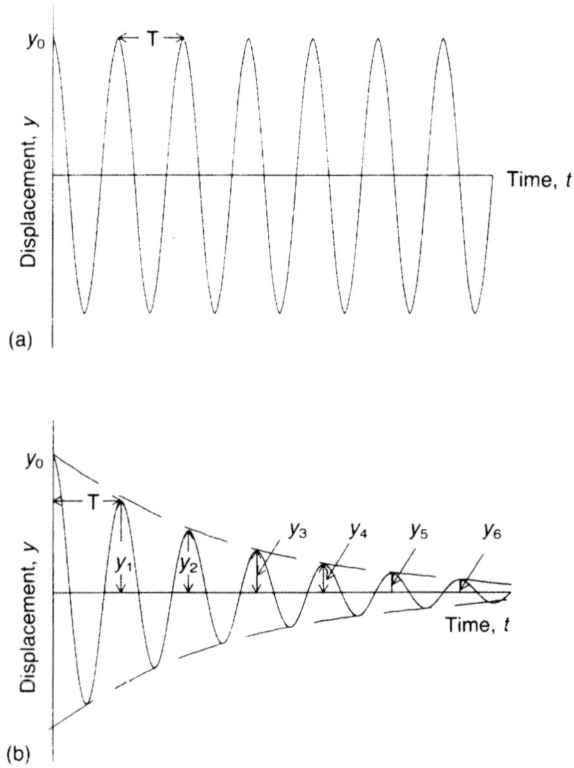

Fig. 2.2 Variation of displacement, y, with time, t, of mass, M, illustrated in Fig. 2.1 when subjected to free vibrations after an initial displacement, y_0, for: (a) zero damping; and (b) finite damping. y_1, y_2, y_3 etc. indicate amplitudes when $t = T$, $2T$, $3T$ etc., and T is the time period. The broken curve indicates the envelope of the decaying sinusoidal curve.

2.1(a) predicts the envelopes actually passing through these points. The heights of the consecutive displacement peaks, i.e. the displacement amplitudes (Fig. 2.2(b)), are respectively $y_0, y_1, y_2 \ldots y_n$ at the times $0, T, 2T \ldots nT$, where n is an integer. The amount of damping per cycle of oscillation is usually expressed by δ, the **logarithmic decrement** (or log. dec.) defined as follows:

$$\exp\delta = y_0/y_1 = y_1/y_2 = y_2/y_3 = \ldots y_{n-1}/y_n = (1/n)y_0/y_n. \qquad (2.3)$$

so that:

$$\delta = \alpha' T = R_m T/2M. \qquad (2.4)$$

It is easy to show that δ is equal to the fractional loss of energy per cycle.

2.2.3 Forced vibrations

When a periodic force $F = F_0 \exp j\omega t$ having a frequency $f = \omega/2\pi$ is applied to the mass in the direction of the axis of the spring, forced vibrations at that frequency are excited and the equation of motion (equation 2.3) takes the following form:

$$\frac{M\mathrm{d}^2y}{\mathrm{d}t^2} + R_\mathrm{m}\mathrm{d}y/\mathrm{d}t + y/C_\mathrm{m} = F_0 \exp j\omega t, \tag{2.5}$$

where $j = (-1)^{1/2}$.

Equation 2.5 is analogous to that for an AC electrical circuit consisting of an inductor, capacitor and resistor in series with an applied electromotive force (EMF). For steady state conditions, i.e. when the amplitudes F_0 of the periodic force and y_0 of the displacement are constant, it can be shown that:

$$u = F/[R_\mathrm{m} + j(\omega M - 1/\omega C_\mathrm{m})], \tag{2.6a}$$

where $u = \mathrm{d}y/\mathrm{d}t$ is the velocity.

Here F is analogous to the EMF and u to the electric current.

Equation 2.6b can be written as:

$$u = F/Z_\mathrm{m}, \tag{2.6b}$$

where Z_m is defined as the **mechanical impedance** as given by:

$$Z_\mathrm{m} = R_\mathrm{m} + j(\omega M - 1/\omega C_\mathrm{m}), \tag{2.7a}$$

$$= R_\mathrm{m} + jX_\mathrm{m}, \tag{2.7b}$$

where X_m is the **mechanical reactance**.

The modulus of Z_m is given by:

$$|Z_\mathrm{m}| = [R_\mathrm{m}^2 + (\omega M - 1/\omega C_\mathrm{m})^2]^{1/2}, \tag{2.7c}$$

and its **phase**, ϕ, by

$$\tan\phi = X_\mathrm{m}/R_\mathrm{m}. \tag{2.7d}$$

The **velocity amplitude**, u_0, is then given by:

$$u_0 = F_0/[R_\mathrm{m}^2 + (\omega M - 1/\omega C_\mathrm{m})^2]^{1/2},$$
$$= F_0/|Z_\mathrm{m}|. \tag{2.8}$$

2.2.4 Resonance

Figure 2.3 shows curves giving the relationships between velocity amplitude u_0 and angular frequency ω for a constant value F_0 of applied force and for different values of R_m; they are called **frequency response curves** and are similar in form to those relating to electrical series LRC circuits (where an inductance, L, a capacitance, C, and a resistance, R,

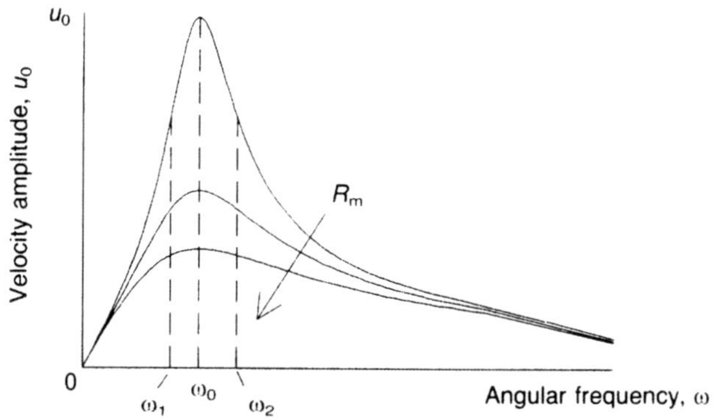

Fig. 2.3 Variations of velocity amplitude, u_0, with angular frequency, ω, for forced vibrations resulting from a periodic force of constant amplitude for different values of quality factor, Q_m. Arrow indicates the effect of increasing R_m. ω_0 indicates the resonant frequency and ω_1 and ω_2 indicate the frequencies at which the maximum velocity amplitude, u_{max}, reduces to, $u_{max}/\sqrt{2}$ for the peak corresponding to minimum damping.

are connected in series). The maximum values of u_0 occur when $\omega = \omega_0 = 1/(MC_m)^{1/2}$, and the mechanical reactance is zero. Here, **resonance** is said to occur and $\omega_0/2\pi$ is called the **resonant frequency**, f_0, which is equal to the fundamental natural frequency for free vibrations as defined in section 2.2.2. The **quality of resonance**, which increases with the height and sharpness of the response curve, is indicated by the **mechanical 'Q' factor**, Q_m, and defined by:

$$Q_m = \omega_0 M/R_m = \pi/\delta, \tag{2.9}$$

where δ is the logarithmic decrement. It can be shown that:

$$Q_m \approx \omega_0/(\omega_2 - \omega_1) = f_0/(f_2 - f_1), \tag{2.10}$$

where ω_1 and ω_2 are the angular frequencies on each side of ω_0 and f_1, f_2 and f_0 are the corresponding frequencies (Fig. 2.3) for which the values of u_0 are equal to the fraction $1/2^{1/2}$ of the maximum value occurring at ω_0, i.e. approximately 3 dB below the maximum (section 2.3.6). The difference $(f_2 - f_1)$ is called the **frequency bandwidth**.

2.2.5 Control of vibrations

Figure 2.3 indicates that the velocity amplitude, u_0, depends on the value of the frequency. At low frequencies, where $\omega \ll \omega_0$, the impedance component ωM is small compared with $1/\omega C_m$ and the value of C_m is predominant. Hence we say that the vibrations are **stiffness controlled**, remembering that stiffness is the reciprocal of compliance.

At high frequencies where $\omega \gg \omega_0$, ωM becomes large compared with $1/\omega C_m$ and the vibrations become **mass or inertia controlled**. At frequencies in the neighbourhood of resonance, ωM and $1/\omega C_m$ tend to cancel out and the mechanical resistance R_m predominates; the vibrations are then **resistance controlled**.

2.3 WAVE MOTION

2.3.1 General considerations

A device which converts energy of one kind to that of another is called a **transducer**. With ultrasonic testing we are concerned with converting electrical energy to acoustic energy and vice versa and the device used is called an **electroacoustic transducer**, which can be either a source or receiver of sound. Most ultrasonic transducers are capable of carrying out both of these conversions with equal efficiency, i.e. they are **reversible** and can thus act simultaneously as transmitters and receivers.

A simple way of understanding the propagation of sound waves is to consider two identical plates, S and R, acting as source and receiver respectively, parallel to one another and separated by an object made from a homogeneous material with which they are in intimate contact (Fig. 2.4). The object is assumed to consist of a number of very thin parallel layers but, for clarity only twenty six of them are shown in the diagram: from A to Z. If the source, S, vibrates in a direction perpendicular to its surface at a given frequency about its mean position, the energy of its vibrations is transmitted in that direction to the adjoining layer A which then oscillates in the same direction at the same frequency. The vibrating layer A then passes on its energy in a similar manner to layer B and progressively to C, D, E etc. which, in turn, vibrate in the same way. The vibrations of the final layer Z are then passed on to the receiver which vibrates in the same direction at

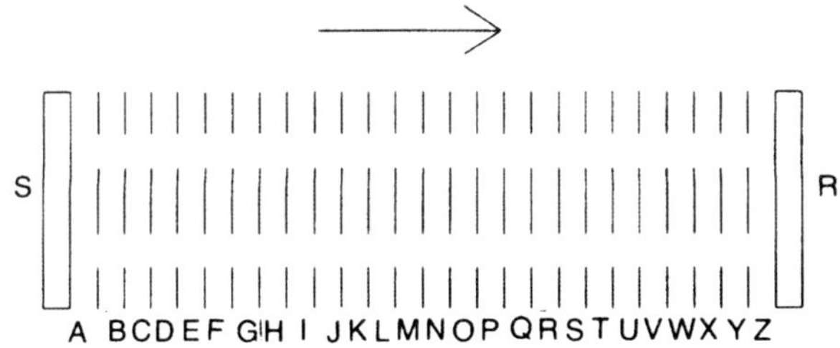

Fig. 2.4 Propagation of sound into a medium divided into thin parallel layers A, B, C, . . . , X, Y and Z. S and R indicate the source and receiver and the single arrow indicates the direction of displacement, x, of sound energy.

the same frequency as that of the source. Because a finite time elapses for these energy transfers to take place, the vibrations of each layer will lag in phase behind those of its predecessor and the waves thus take a finite time to pass from the source to the receiver. The leading surface of the waves is called the **wave front**.

This model illustrates the method of propagation of compression waves as defined in section 2.1. The same model can be employed to show the radiation of shear waves as a result of the source vibrating in a direction tangentially to its surface. It should be thoroughly understood that the actual layers and even the particles (e.g. atoms or molecules) forming them do not progress from the source to the receiver. It is only the energy which moves in this way.

In this model the propagation of **plane waves** is discussed. Here the phase of vibration within any plane parallel to the surface of a plane source is constant, i.e. the wave front is plane. If the source is a pulsating sphere, **spherical waves** are propagated and the source may be regarded conveniently as a point situated at the centre of the sphere. At distances remote from a spherical source where the curvature of the wave fronts becomes small, spherical waves approximate to plane waves.

2.3.2 Equations of motion of plane waves

For a source of plane waves vibrating with simple harmonic motion at constant amplitude y_0 and frequency $f = \omega/2\pi$, it can be assumed that the particles of the medium of propagation in contact with it are vibrating in an identical manner. Thus the particle displacement, y, from the mean position at zero distance from the source, i.e. $x = 0$, can be expressed by:

$$y = y_0 \cos \omega t \quad \text{(cf equation 2.1b).} \tag{2.11}$$

y_0 is the **particle displacement amplitude**. Consider plane waves travelling in the x-direction with a constant speed c. The transit time t' of the waves to travel a distance x from the source is then equal to x/c. The phase of the vibrations at this position thus lags behind that where $x = 0$ by an amount $\omega t'$ and y is then expressed as:

$$y = y_0 \cos \omega(t - t'), \quad \text{i.e.}$$

$$y = y_0 \cos \omega(t - x/c). \tag{2.12a}$$

For a single time period T, during which the phase angle has changed by 2π, the waves travel by a distance $\lambda = cT$, where λ is known as the **wavelength**.

Thus:

$$y = y_0 \cos 2\pi(t/T - x/\lambda), \tag{2.12b}$$

$$y = y_0 \cos(\omega t - kx), \tag{2.12c}$$

where $k = 2\pi/\lambda = \omega/c$ is the **wave number**. It is thus seen that the speed of sound c is equal to λ/T, i.e. λf.

The **particle velocity**, u, is given by the partial differential $\partial y/\partial t$ from which it can be seen that by differentiating equation 2.12c with respect to t:

$$u = -\omega y_0 \sin(\omega t - kx) = u_0 \cos(\omega t - kx + \pi/2), \tag{2.12d}$$

where u_0 is the **particle velocity amplitude** and from which it can be seen that the phase of u leads that of y by 90°. It is usual to zero the scales for x and t in such a way that the phase functions can be expressed simply as either:

$$\cos(\omega t - kx), \quad \sin(\omega t - kx) \quad \text{or} \quad \exp j(\omega t - kx),$$

where $j = (-1)^{1/2}$, although it is common practice to express the variations of u and the acoustic pressure p (see below) as functions of $\sin(\omega t - kx)$. The use of the exponential has the advantage of simplifying differentiations and integrations. Thus if a function Φ represents either y, u or p, we can write:

$$\Phi = \Phi_0 \exp j(\omega t - kx). \tag{2.13}$$

If equation 2.13 is differentiated twice with respect to t and k, we have:

$$\frac{\partial^2 \Phi}{\partial x^2} = -k^2 \Phi_0 \exp j(\omega t - kx),$$

$$\frac{\partial^2 \Phi}{\partial t^2} = -\omega^2 \Phi_0 \exp j(\omega t - kx),$$

thus giving:

$$\frac{\partial^2 \Phi}{\partial t^2} = c^2 \partial^2 \Phi / \partial x^2 \tag{2.14a}$$

since $c = \omega/k$. This is called the **general wave equation for plane waves**. Φ represents any wave characteristic such as particle displacement y, particle velocity u and **acoustic pressure** p: p is defined by the excess pressure exerted by the sound waves, at a given point, which gives rise to the strain. For shear waves, the term **acoustic stress** is clearly more appropriate. Taking into account the x, y and z directions, equation 2.14a can be written more generally as:

$$\frac{\partial^2 \Phi}{\partial t^2} = c^2 \left(\frac{\partial^2 \Phi}{\partial x^2} + \frac{\partial^2 \Phi}{\partial y^2} + \frac{\partial^2 \Phi}{\partial z^2} \right),$$

$$\frac{\partial^2 \Phi}{\partial t^2} = c^2 \nabla^2 \Phi. \tag{2.14b}$$

2.3.3 The speed of plane waves

Consider plane compression waves propagated in the x-direction in a sample of homogeneous material of uniform density and let AB (Fig. 2.5) represent a thin parallel-sided layer, forming part of the material, with the respective surfaces A and B at right angles to the direction of propagation and at distances x and $x + \delta x$ from some convenient origin (e.g. the source). At some time, t, the wave motion gives rise to a displacement of the surface A by an amount y to a position A′ so that its distance from the origin is increased to $y + x$ and the surface B moves to a position B′. Assuming the displacements to be small, Taylor's theorem predicts that the distance BB′ is equal to $y + (\partial y/\partial x)\delta x$ to a first-order approximation.

Taking a unit area of cross-section, the mechanical strain suffered by the layer can be seen to equal $\partial y/\partial x$. The acoustic pressure, p, is evaluated from Hooke's law, i.e.:

$$p = -q\frac{\partial y}{\partial x} \tag{2.15}$$

where q is the appropriate modulus of elasticity. The negative sign arises because the pressure represents a compressive stress. A characteristic of acoustic waves is that the variations in strain occur so rapidly that there is no time for the system to settle down to a state of thermal equilibrium, i.e. adiabatic conditions are assumed and the appropriate elastic modulus is an adiabatic one, as opposed to an isothermal modulus, appertaining to static conditions. For gases, the difference between the isothermal and adiabatic moduli is considerable but, for solids and liquids, it can be taken as equal to zero except for high precision work appropriate to research and standards laboratories.

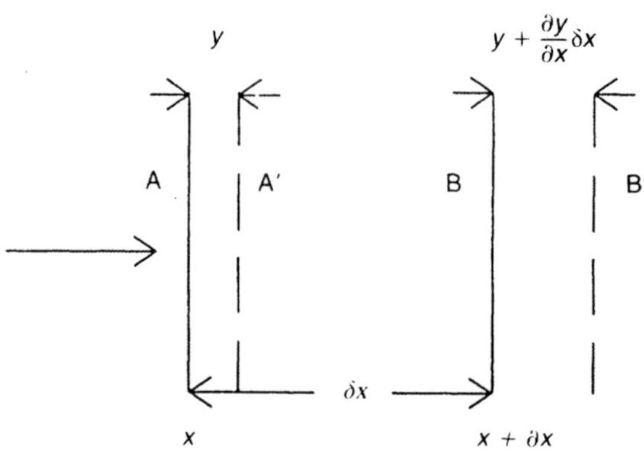

Fig. 2.5 Strain of layer AB of thickness, δx, of homogeneous material caused by passage of longitudinal plane waves. A′B′ indicates displaced layer. Arrow indicates direction of waves.

Consider now a layer CD, similar to AB, in the material and let the values of acoustic pressure be p and $p + (\partial p/\partial x)\delta x$, at C and D, respectively. From Newton's third law of motion, equal and opposite pressures act on each of these boundaries with the result that the layer is compressed by $+p$ at C and $-[p + (\partial p/\partial x)\delta x]$ at D. This results in an unbalanced pressure $-(\partial p/\partial x)\delta x$ resulting an acceleration $\partial^2 y/\partial t^2$ of the layer in accordance with Newton's second law of motion. Thus the force per unit area, i.e. pressure, is equal to $\rho\delta x\partial^2 y/\partial t^2$, where ρ is the density of the material. $\rho\delta x$ is thus the mass per unit area of the layer. Hence:

$$\partial p/\partial x = -\rho\partial^2 y/\partial t^2. \tag{2.16}$$

Combining equations 2.15 and 2.16 gives:

$$\partial^2 y/\partial t^2 = (q/\rho)\partial^2 y/\partial x^2, \tag{2.17}$$

and from equations 2.14a and 2.17 we obtain:

$$c = (q/\rho)^{1/2}. \tag{2.18}$$

The same equation can be derived for shear wave propagation for which the source is subjected to periodic shear stresses and thus perform transverse vibrations in direction of the shear. It is shown later (section 3.2) how this equation is modified to suit the different types of wave propagation.

2.3.4 Specific acoustic impedance

In section 2.2 the expression **mechanical impedance** is shown to be analogous to electrical impedance. A similar analogy exists with respect to electrical and acoustic vibrations, i.e. acoustic pressure p with EMF (and electrical potential difference), particle displacement y with charge and particle velocity u with current. The acoustic analogy for electrical impedance is the **specific acoustic impedance**, Z_a, i.e.:

$$Z_a = p/u. \tag{2.19}$$

It is equal to the mechanical impedance Z_m per unit area of cross-section of the sound beam and it also has two components, i.e.

$$Z_a = R_a + jX_a \tag{2.20}$$

where $j = (-1)^{1/2}$. The real component R_a is called the **characteristic impedance**.

The reader should not confuse the terms specific acoustic impedance and characteristic impedance with acoustic impedance. The latter is commonly used in audioacoustics and has an entirely different meaning: the ratio of Z_m to the square of the area of cross-section.

Considering plane progressive waves in a non-absorbent material, equation 2.15 and the equations 2.12 show that:

$$p = -q\frac{\partial y}{\partial x} = -qky_0 \sin(\omega t - kx),$$

and since:

$$u = \frac{\partial y}{\partial t} = -\omega y_0 \sin(\omega t - kx),$$

we have, for all values of x and t:

$$Z_a = p/u = qk/\omega = \rho c = R_a, \tag{2.21}$$

since $c = \lambda f$, $k = 2\pi/\lambda$ and $\omega = 2\pi f$.

Because there is no reactive component, Z_a is entirely real and equal to R_a, i.e. ρc, for **plane progressive waves** in a non-absorbent material. This would not be the case for plane progressive waves in an absorbent material (Appendix C) or for spherical waves and stationary plane waves, for which the expressions for y, u and p are more complicated than those shown in the equations 2.13 (for example, Kinsler *et al.* 1982). Equation 2.21 shows that the unit of R_a is the $\mathrm{kg\,m^{-2}\,s^{-1}}$ which is commonly expressed as the **rayl**.

2.3.5 Acoustic intensity

The flow of acoustic energy per unit area in a sound field is expressed as the **acoustic intensity**. If we consider a small plane surface of area δA at right-angles to the direction of propagation, through which the rate of flow of energy is δW, the ratio $\delta W/\delta A$ is defined the average acoustic intensity within that area. As δA tends to zero, the surface reduces to a point so that $\delta W/\delta A$ approaches the differential $\mathrm{d}W/\mathrm{d}A$, i.e. the acoustic intensity I at that point is given by:

$$I = \frac{\mathrm{d}W}{\mathrm{d}A}. \tag{2.22}$$

For spherical waves, the value of I reduces with distance r from the source. Let W be the acoustic power, i.e. rate of change of energy, of a point source radiating uniformly in all directions in a uniform isotropic and non-absorbent medium. It is easy to see that the intensity is the same at all points on a spherical surface of radius r with the source at its centre. Because the area of the spherical surface is equal to $4\pi r^2$, the value of the intensity at a point at a distance r from the source is given by:

$$I = W/4\pi r^2. \tag{2.23}$$

The intensity thus varies in accordance with the inverse square law: I is proportional to $1/r^2$.

For plane waves in a uniform non-absorbent material, the principle of conservation of energy shows that the intensity I remains the same at all points in the sound field, irrespective of distance. Let E represent the amount of sound energy contained in unit volume of the field: the **energy density**. This energy travels through each unit cross-section of the field with a velocity equal to the speed of sound, c. The length of a column of such a cross-section and, hence the volume, per unit time are thus numerically equal to c and the energy contained by the column is equal to cE which, by definition, is equal to the intensity I:

$$I = cE. \tag{2.24}$$

Consider now the mass, M, of a single vibrating particle such as a molecule, in the medium. The vibrations of this particle can be treated as being analogous to those of the mass and spring model described in section 2.2 where the vibrational energy E_m is given by equation A.1. This equation may be expressed as:

$$E_m = \tfrac{1}{2}Mu^2 + \tfrac{1}{2}y^2/C_m. \tag{2.25}$$

For a constant output of the source, E_m stays the same at all times and if we select a value of time, t, corresponding to $y = 0$, equations 2.12 and 2.13 show that $u = u_0$ for which $E_m = \tfrac{1}{2}Mu_0^2$. Taking unit volume of the material, it is seen that the numerical value of M is equal to that of the density ρ, i.e. E_m is numerically equal to the energy density E. Hence,

$$E = \tfrac{1}{2}\rho u_0^2, \tag{2.26}$$

so that

$$I = cE = \tfrac{1}{2}\rho c u_0^2 = \tfrac{1}{2}R_a u_0^2, \tag{2.27}$$

since $R_a = \rho c$ (equation 2.21). This equation is analogous to that relating electrical AC power P to the current amplitude I_0 and resistance R:

$$P = \tfrac{1}{2}I_0^2 R.$$

Equation 2.27 can be expressed alternatively as:

$$I = \tfrac{1}{2}p_0^2/\rho c \quad (\text{cf } P = \tfrac{1}{2}V_0^2/R), \tag{2.28}$$

where V_0 is the peak voltage, or

$$I = \tfrac{1}{2}p_0 u_0 \quad (\text{cf } P = \tfrac{1}{2}V_0 I_0) \tag{2.29}$$

As with electrical quantities, the root mean square (r.m.s.) values can be substituted for the amplitudes, thus eliminating the factor ½.

2.3.6 The decibel scale

The equations 2.27 to 2.29 show that the intensity varies with the square of the amplitude. In practice, variations of intensity and amplitude are expressed in terms of the **decibel** (dB), as follows:

$$\text{increase in level} = 10\log_{10}(I/I_0) = 20\log(A/A_0)\text{dB} \qquad (2.30)$$

where A represents either P_0, u_0 or y_0 and the subscripts 0 to I and A indicate a reference level. It can be seen that the same value in dB is obtained for either intensity or pressure levels. With most ultrasonic tests, electrical voltage outputs are observed; because of the analogy, they can be shown to be directly proportional to acoustic pressures (equation 4.4a).

2.3.7 Pressure of radiation

Sound waves consist of energy in motion and, hence, exert a steady pressure on any object placed in an acoustic field. By definition, acoustic intensity is equal to the power (i.e. energy per unit time) acting on and perpendicular to unit area. Furthermore, power is equal to the product of force and speed. Consequently, the intensity is the product of the speed of the waves and this steady pressure, the **pressure of radiation** P_r:

$$I = P_r c \qquad (2.31)$$

Comparing equations 2.24 and 2.31 shows that P_r is equal to the energy density.

2.4 ATTENUATION OF PLANE WAVES

2.4.1 General consideration

Attenuation of plane acoustic waves arises mainly from the following causes:

- deviation of energy from the parallel beam by reflection, refraction, diffraction and scattering; and
- absorption, for which some of the sound energy is converted into heat by one of a number of causes.

The phenomena of reflection, refraction and diffraction take place as a result of discontinuities and are discussed later in this Chapter. Scattering and absorption losses are characteristic of the material of the object being tested and their evaluation can yield valuable information about the structure and physical properties of the medium. These phenomena are discussed in section 3.3.

2.4.2 Attenuation coefficient

Consider a parallel-sided layer of infinitesimal thickness dx at a distance x from a source of plane acoustic waves. For uniform attenuation, 2α the fractional loss of energy per unit path-length is constant and for a path-length dx, we have:

$$dE/E = -2\alpha dx, \tag{2.32}$$

where E is the energy density. Since the intensity, I, is proportional to E:

$$dI/I = -2\alpha dx. \tag{2.33}$$

Putting $I = I_0$ when $x = 0$ and integrating with respect to x yields:

$$I = I_0 \exp(-2\alpha x), \tag{2.34a}$$

and, because I is proportional to the square of the amplitude, A,

$$A = A_0 \exp(-\alpha x), \tag{2.34b}$$

where $A = A_0$ when $x = 0$. Expressing equation 2.13 in terms of acoustic pressure, p, and allowing for attenuation, we have:

$$p = p_0 \exp(-\alpha x) \exp j(\omega t - kx). \tag{2.35}$$

The constant, α, is defined as the **attenuation coefficient** for the material and may be expressed in the unit neper, m^{-1}. Alternatively, attenuation can be expressed in terms of the amplitude loss per cycle, or per wavelength as given by $\alpha\lambda$ which can be seen to equal to the logarithmuic decrement, δ, (equation 2.4). However, it is common practice nowadays to use the decibel scale, i.e. 1 neper is equivalent to $10 \log_{10} e = 8.686\,dB$.

2.5 REFLECTION AND TRANSMISSION OF PLANE WAVES

2.5.1 Reflection and transmission at normal incidence

Consider a beam of plane waves, either compression or shear waves, incident normally at a plane boundary separating two 'semi-infinite' isotropic homogeneous materials, 1 and 2, having characteristic impedances R_1 and R_2, respectively (Fig. 2.6). In general, part of the incident wave energy in material 1 is reflected back along the incident path and the remainder is transmitted through material 2. Let p_i, p_r and p_t be the values of acoustic pressure (or stress) for the incident, reflected and transmitted waves, respectively and u_i, u_r and u_t the corresponding values of particle velocity. For non-absorbent media, the wave equations for acoustic pressure and particle velocity can be written as:

$$p_i = A_1 \sin(\omega t - k_1 x) = u_i R_1, \tag{2.36}$$

$$p_r = B_1 \sin(\omega t + k_1 x) = -u_r R_1, \tag{2.37}$$

$$P_t = A_2 \sin(\omega t - k_2 x) = u_t R_2, \tag{2.38}$$

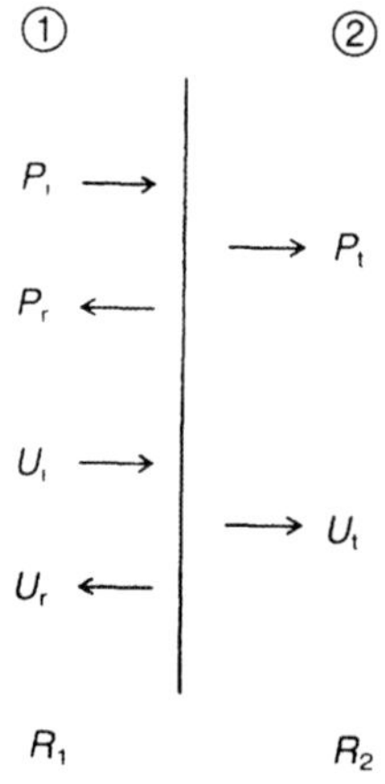

Fig. 2.6 Acoustic reflection and transmission for normal incidence at a plane boundary between different material media 1 and 2 of characteristic impedances R_1 and R_2, respectively. p_i, p_r and p_t represent the acoustic pressures and u_i, u_r and u_t the particle velocities for incident, reflected and transmitted waves, respectively.

in accordance with equation 2.13. A_1, B_1 and A_2 are the pressure amplitudes and k_1 and k_2 are the wave numbers, $2\pi/\lambda$, for the materials 1 and 2, respectively: x is positive when its direction is that of the incident beam. The reversals in sign shown in equation 2.37 arise because the reflected beam travels in the negative direction of x and the velocity of sound becomes $-c$.

The following boundary conditions apply at all times.

- The pressures should be the same on both sides of the boundary to preserve continuity:

$$p_t = p_i + p_r.$$

- The particle velocities normal to the boundary should be the same on both sides to enable the two materials to remain in contact with one another:

$$u_t = u_i + u_r.$$

Putting $x = 0$ at the boundaries, these two equations become:

$$A_2 = A_1 + B_1$$

and

$$R_1 A_2 = R_2(A_1 - B_1)$$

Thus
$$p_t/p_i = A_2/A_1 = 2R_2/(R_1 + R_2) \tag{2.39}$$

$$p_r/p_i = B_1/A_1 = (R_2 - R_1)/(R_1 + R_2). \tag{2.40}$$

The ratio $r = B_1/A_1$ defines the **amplitude reflection coefficient**. Also:

$$u_t/u_i = A_2/R_1/A_1R_2 = 2R_1/(R_1 + R_2), \quad \text{and} \tag{2.41}$$

$$u_r/u_i = -B_1/A_1 = (R_1 - R_2)/(R_1 + R_2). \tag{2.42}$$

When R_2 is greater than R_1, i.e. when the waves pass from an acoustically less dense material to a denser one (e.g. from air to a solid), equation 2.40 shows that there is no change of phase of acoustic pressure on reflection. On the other hand, equation 2.42 indicates a 180° phase change of particle velocity (and, hence, particle displacement) as a result of the negative sign appended to the ratio B_1/A_1.

At the boundary between the two materials we can define a quantity called the **reflection coefficient**, α_r, which is defined as the ratio of the intensity, I_r, of the reflected waves to that, I_i, of the incident waves. α_r is seen to equal r^2, where r is the amplitude reflection coefficient (equation 2.40). Another factor, called the **transmission coefficient** α_t, is defined as the ratio of the intensity, I_t, of the transmitted waves to I_i. Assuming no energy losses at the boundary, we see that

$$I_i = I_r + I_t,$$

so that

$$\alpha_r + \alpha_t = 1. \tag{2.43}$$

Because the intensity is proportional to the square of the acoustic pressure amplitude (equation 2.28) in a given material, it is seen that

$$\alpha_r = I_r/I_i = B_1{}^2/A_1{}^2 = (R_2 - R_1)^2/(R_1 + R_2)^2, \tag{2.44}$$

and from equation 2.43, we have

$$\alpha_t = 1 - \alpha_r = 4R_1R_2/(R_1 + R_2)^2. \tag{2.45}$$

The transmission coefficient may be expressed either as a fraction, as a percentage or in decibels (section 2.3.6), $-10\log_{10}\alpha_t$, and may be referred to as the **transmission loss**. Thus a coefficient of 10% is approximately equivalent to a loss of 10 dB and one of 80% to a loss of 1 dB. The term **degree of coupling** is often used. This is high for a high transmission coefficient, i.e. low loss, and low for a low transmission coefficient, i.e. high loss. It should be noted that, if we define an amplitude transmission coefficient using equation 2.39, it would have a value greater than unity when R_2 is greater than R_1, even though the intensity amplitude coefficient defined above is less than unity.

The equations 2.44 and 2.45 show that when $R_1 = R_2$, α_t is at its maximum value of 1 and $\alpha_r = 0$. Ideally, this gives rise to perfect coupling between the two materials. In practice, this is rarely achieved but very good coupling arises when R_1 and R_2 have values of the same order of magnitude, i.e. for which α_t is greater than 70% and thus a transmission loss of less than about 1.5 dB. Coupling is poor between a solid and a gas, such as air, when α_t is about 0.005–0.006 % with a corresponding transmission loss of about 43 dB, which is unacceptable for ultrasonic testing. When a solid is in contact with a liquid, α_t is about 20–25 %, a loss of 6–7 dB, which is usually acceptable for equipment having a dynamic range of more than 100 dB. Values of characteristic impedance applicable to compression waves in a number of commonly used materials are provided in Table 2.1, from which the reader can calculate values of α_t and transmission loss.

Conditions for reflection and transmission may be altered considerably if a parallel layer of an intervening substance of characteristic impedance, R_0, is located between the two materials. It can be shown (see, for example Kinsler *et al.* 1982) that:

$$\alpha_t = 4R_1R_2/[(R_1 + R_2)^2 \cos^2 kl + (R_0 + R_1R_2/R_0)^2 \sin^2 kl], \quad (2.46)$$

where l is the thickness and $k = 2\pi/\lambda$ is the wave number for the third medium.

Three special cases are of interest.

- When the intervening material is very thin compared with the wavelength, $kl \ll 1$, for which $\cos^2 kl$ tends to unity and $\sin^2 kl$ becomes negligible. Provided that R_0 is not too small compared with R_1 and R_2, equation 2.46 approximates to equation 2.45 and the nature of the intervening material has little effect on α_t. On the other hand, if the value of R_0 is small compared with R_1 and R_2, the value

Table 2.1 Approximate values of characteristic impedances for compression waves in some commonly used materials in bulk form at room temperature. The data given comprise typical values. Exact values depend on structure, previous history, impurity content and physical conditions

Material	*Speed of sound, c* ($m\,s^{-1}$)	*Density,* ρ ($kg\,m^{-3}$)	*Characteristic impedance,* $\rho c = R_a$ *(rayl)*
Aluminium	6400	2700	1.7×10^7
Brass	4400	8400	3.7×10^7
Copper	4800	8900	4.3×10^7
Gold	3200	20 000	6.4×10^7
Lead	2200	11 400	2.5×10^7
Mild steel	5500–6000	7800	$4.3–4.7 \times 10^7$
Nickel	5600	8900	5.0×10^7
Perspex	2700	1200	3.2×10^6
Lead zirconate titanate (PZT)	4800	7500	3.6×10^7
Lubricating oil	1400–1800	900	$1.3–1.6 \times 10^6$
Water	1500	1000	1.5×10^6
Air	330	1.3	430

of α_t can be very low. Consider the transmission of longitudinal waves from a lead zirconate titanate (PZT) transducer to a steel sample. Because of inherent surface roughness, there is no proper contact between the two materials and we have to consider a thin layer of air between them. Assuming an average layer thickness of 10 μm, characteristic impedances, in rayls, of PZT, steel and air are 3.6×10^7, 4.7×10^7 and 430, respectively and the speed of sound in air equal to 330 m s^{-1}, it can be seen from equation 2.46 that the value of α_t, for a frequency of 1 MHz, is 1.2×10^{-8} which corresponds to a loss of 79 dB. If, however, the surfaces are coupled with a layer of lubricating oil, for which the speed of sound and characteristic impedance are 1700 m s^{-1} and 1.5×10^6 rayl, respectively, the value of α_t increases to 0.78 which corresponds to a loss of only about 1 dB. This is very close to the loss value of 0.3 dB (α_t = 0.94) obtained from equation 2.45 which assumes perfect coupling between the two solid surfaces.

- When the thickness of the intervening material is exactly equal to an integral number of half-wavelengths, $\cos kl = 1$ and $\sin kl = 0$. corresponding to a resonance frequency (section 2.6.5). Equation 2.46 thus becomes identical to equation 2.45. This is unlikely to happen in practice because of the operation of transducers over finite bands of frequency. However, for narrow-band sources, a considerable increase in transmission does occur at one of the resonance frequencies.
- When the thickness of the intervening medium is equal to an odd number of quarter-wavelengths, $\cos kl = 0$ and $\sin kl = 1$, hence

$$\alpha_t = 4R_1R_2/(R_0 + R_1R_2/R_0)^2. \tag{2.47}$$

100 % transmission then occurs when $R_0^2 = R_1R_2$. In practice, transmission losses are considerably reduced if the value of R_0 lies between those of R_1 and R_2, even when the thickness of the intervening layer is not exactly equal to an odd number of quarter wavelengths. Consider, for example, the coupling between a ceramic transducer and water. Equation 2.45 and Table 2.1 show that, for direct contact, α_t = 24 %, and the transmission loss is 6.2 dB. On interposing a layer of hard rubber of quarter-wavelength thickness ($R_0 = 2.9 \times 10^6$ rayls and l = 0.6 mm at a frequency of 1 MHz) equation 2.47 shows that the value of α_t increases to 65% with a reduction of transmission loss to 2 dB. This equation shows that α_t is equal to 100 % when $R_0^2 = R_1R_2$, which is an ideal case.

Although, in theory, the subject matter discussed above is applicable to shear-wave propagation, difficulties arise in practice resulting from the need to introduce a fluid couplant for transmission between two solids. It is, of course, possible to overcome this difficulty by using an adhesive couplant provided that it can be removed after testing without interfering with the properties of the object being examined.

2.5.2 Reflection and transmission at oblique incidence

Consider now a parallel beam XO of longitudinal plane waves incident obliquely at a plane boundary PQ separating the materials 1 and 2 (Fig. 2.7), where O is the point of incidence at the boundary, OY and OZ are, respectively, the reflected and transmitted (i.e. refracted) longitudinal beams. When one or both of the materials supports a shear stress, mode conversion takes place at the boundary. The stress exerted by an oblique beam having components resolved perpendicular to and parallel with that boundary. Hence the layers of the media in the vicinity of the boundary suffer both compressive and shear stresses. Hence shear waves, where appropriate, and compression waves may be reflected and transmitted as indicated by OY′ and OZ′, respectively.

The speeds and directions of the sound associated by these four beams are determined by Snell's law:

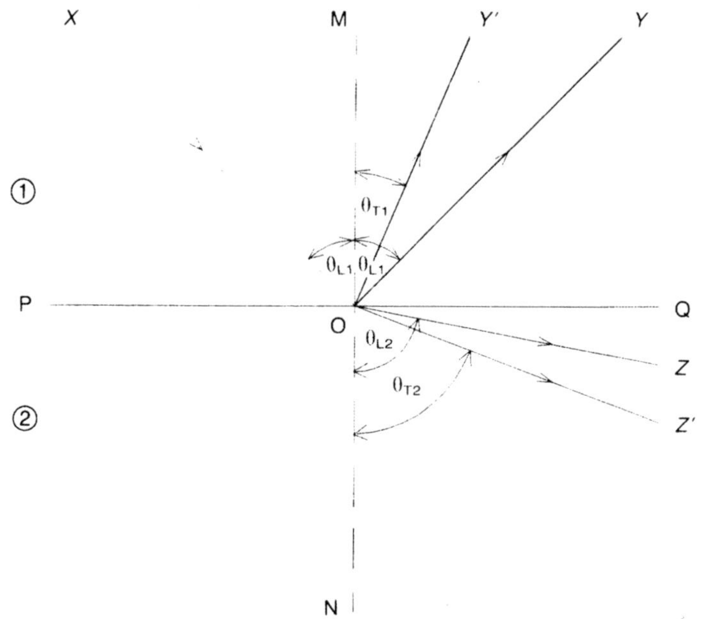

Fig. 2.7 Reflection and transmission of sound for oblique incidence of compression waves at a plane boundary PQ between two different media 1 and 2. θ_{L1}, θ_{T1}, θ_{L2} and θ_{T2} are the angles between the normal to the boundary for the incident and reflected compression waves, the reflected shear waves, the transmitted compression waves and the transmitted shear waves, respectively. MON represents the normal to the boundary PQ through the point of incidence.

$$c_{L1}/\sin\theta_{L1} = c_{T1}/\sin\theta_{T1} = c_{L2}/\sin\theta_{L2} = c_{T2}/\sin\theta_{T2} \qquad (2.48)$$

where c_{L1} and c_{T1} represent the compression and shear wave velocities in the medium of incidence 1 and c_{L2} and c_{T2} in the medium of transmission 2, respectively. θ_{L1}, θ_{T1}, θ_{L2} and θ_{T2} are the angles between the normal to the boundary and the corresponding beams, as indicated in Fig. 2.7. Because the shear modulus of elasticity is always less than that associated with compression and tensile stresses (section 3.2), the speed of shear waves must be less than that of compression waves.

If medium 1 is a fluid and medium 2 a solid, with c_{L1} being less than both c_{L2} and c_{T2}, equation 2.48 shows that both of the transmitted rays are refracted away from the normal, with the angle θ_{L2} being greater than θ_{T2} for the reason given above. Let the angle θ_{L1} be increased to a value θ_{c1} for which θ_{L2} is equal to 90° (Fig. 2.8(a)). θ_{c1} is called the **first critical angle** and, for angles of incidence greater than this, only shear waves enter medium 2. This is indicated by an abrupt increase in intensity of the reflected waves, which are compression waves. A further increase in θ_{L1} leads to a second critical angle θ_{c2} for which θ_{T2}

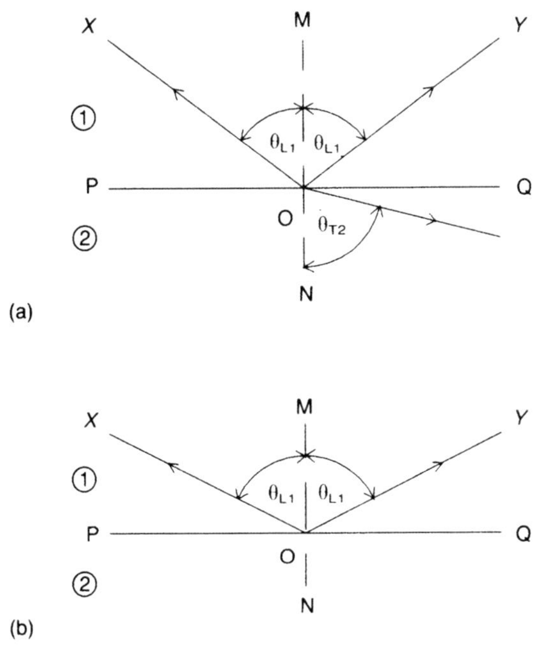

Fig. 2.8 Effect of oblique incidence of compression waves at a plane boundary between two different media 1 and 2, when R_1 is less than R_2 for: (a) the first critical angle, θ_{c1}; and (b) the second critical angle, θ_{c2}. θ_{L1} and θ_{T2} represent the angle between the normal MON and the boundary PQ for the incident and reflected compression waves and the transmitted shear waves.

= 90° and with a corresponding abrupt increase in the intensity of the reflected beam. An even further increase in θ_{L1} leads to a third critical angle θ_{c3} for which **Rayleigh**, i.e. **surface waves** are propagated along the boundary (section 3.5); this is indicated by a **decrease** in the intensity of the reflected waves. The speed c_R of the surface waves is approximately equal to $0.9c_{T2}$. It is seen from equation 2.48 that:

$$c_{L2} = c_{L1}/\sin\theta_{c1}, \tag{2.49a}$$

$$c_{T2} = c_{L2}/\sin\theta_{c2}, \quad \text{and} \tag{2.49b}$$

$$c_R = c_{L1}/\sin\theta_{c3}. \tag{2.49c}$$

If medium 1 is a solid and medium 2 a fluid, mode conversion with the internally reflected beam occurs but only longitudinal waves are transmitted. When, as is usual, c_{L1} and c_{T1} are greater than c_{L2}, the angles θ_{L1} and θ_{T1} are both greater than θ_{L2} and the phenomena described in the previous paragraph do not arise. However, because of the large mismatch of characteristic impedance between the materials, especially if medium 2 is air and medium 1 a metal, most of the sound energy is reflected back into medium 1 in the form of both compression and shear waves. This phenomenon is discussed in more detail in sections 5.3.2 and Appendix B, which deals with mode conversion signals obtained with the testing of rods using the pulse-echo method.

The values of the transmission and reflection coefficients for oblique incidence are dependent on the characteristic impedances of the media in relation to the type of wave, i.e. compression, shear etc., and the angles of incidence. The study of this phenomenon is somewhat difficult and the interested reader is referred to Mayer (1965) for further information. An important application is the use of the ultrasonic goniometer (section 5.8) for the measurement of the speeds of both compression and shear waves in highly attenuating solids and in solid objects having only a single plane surface. With this method, the object is located in water through which an acoustic beam is propagated so as to be incident at an angle θ_{L1} (Fig. 2.7). Mayer has indicated that the values of the reflection coefficient α_r, the compression-wave transmission coefficient α_{tL} and the shear-wave transmission coefficient α_{tT} for a given angle of incidence θ_{L1} are given by:

$$\alpha_r = [1 - C(1 - 2A)]^2/[1 + C(1 - 2A)]^2, \tag{2.50a}$$

$$\alpha_{tL} = 4BC/[1 + C(1 - 2A)]^2, \quad \text{and} \tag{2.50b}$$

$$\alpha_{tT} = 4D(\rho_2/\rho_1)/[1 + C(1 - 2A)]^2, \tag{2.50c}$$

where

$$A = \sin\theta_{T2}\sin 2\theta_{T2}[\cos\theta_{T2} - (c_{T2}/c_{L2})\cos\theta_{L2}],$$

$$B = (\cos 2\theta_T)^2,$$

$$C = (\rho_2 c_{T2} \cos \theta_{L1})/(\rho_1 c_{L1} \cos \theta_{L2}),$$

$$D = (c_{T2}/c_{L1})^2 \sin 2\theta_{1L} \sin 2\theta_{2T},$$

and ρ_1 and ρ_2 represent the respective densities of the media. Figure 2.9 illustrates the relationship between (a) α_r and θ_{L1} and (b) α_t and θ_{L1} for a Perspex–water boundary for incident compression waves in water.

2.6 STATIONARY WAVES

2.6.1 General considerations

With the reflection of plane waves at normal incidence to a plane boundary (section 2.5.1), the incident and reflected beams in medium

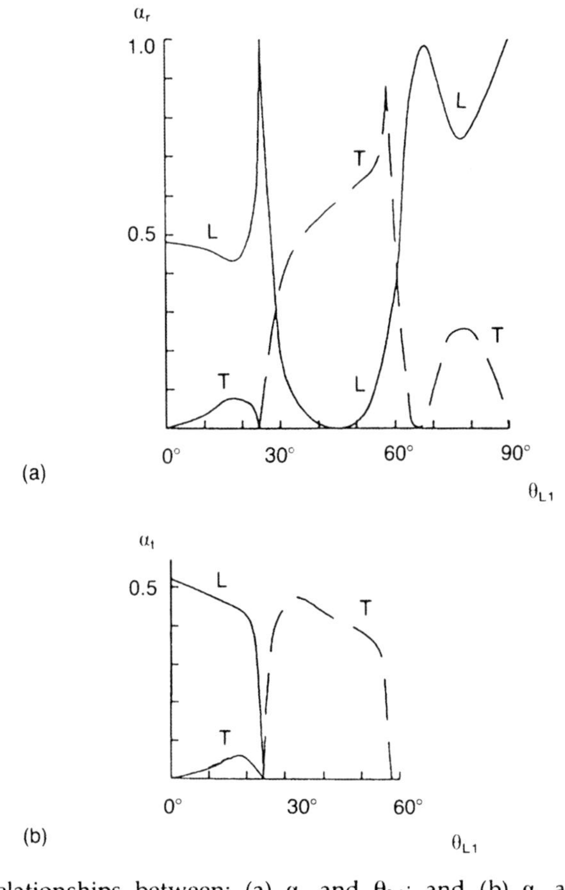

Fig. 2.9 Relationships between: (a) α_r and θ_{L1}; and (b) α_t and θ_{L1} for a Perspex–water boundary (L = compression waves, T = shear waves).

1, i.e. the medium of incidence, interfere with one another to form **stationary** or **standing** waves. The resultant acoustic pressure, p, and particle velocity, u, at any point in this medium are given by:

$$p = p_i + p_r \quad \text{and} \quad u = u_i + u_r,$$

where the subscripts i and r refer to incident and reflected waves, respectively. Assuming that the medium is non-absorptive we have, from equations 2.36 and 2.37,

$$\begin{aligned} p &= p_0 \sin(\omega t - kx) + rp_0 \sin(\omega t + kx) \\ &= rp_0[\sin(\omega t - kx) - \sin(\omega t + kx)] + (1 - r)p_0 \sin(\omega t - kx), \end{aligned} \tag{2.51a}$$

and

$$\begin{aligned} u &= u_0 \sin(\omega t - kx) - ru_0 \sin(\omega t + kx) \\ &= ru_0[\sin(\omega t - kx) - \sin(\omega t + kx)] + (1 - r)u_0 \sin(\omega t - kx), \end{aligned} \tag{2.51b}$$

p_0 and u_0, respectively, are the pressure and particle velocity amplitudes for the incident waves, and r the amplitude reflection coefficient (equation 2.40), as seen by substituting $A_1 = p_0 = R_1 u_0$ and $B_1 = rp_0 = -rR_1 u_0$.

The following examples are now considered.

2.6.2 R_2 is greater than R_1

Equations 2.40 and 2.42 show that, on reflection, there is no change in phase for p but a 180° change for u. Hence in equation 2.51a, r is positive, $+|r|$, and in equation 2.51b, it is negative, $-|r|$. By trigonometrical manipulation they can then be expressed as:

$$p = 2|r|p_0 \cos kx \sin \omega t + (1 - |r|)p_0 \sin(\omega t - kx), \tag{2.52a}$$

$$u = 2|r|u_0 \sin kx \cos \omega t + (1 - |r|)u_0 \sin(\omega t - kx). \tag{2.52b}$$

The resultant amplitudes P and U of p and u, respectively, thus vary with x,

$$P = 2|r|p_0 \cos kx + (1 - |r|)p_0, \tag{2.52c}$$

$$U = 2|r|u_0 \sin kx + (1 - |r|)u_0. \tag{2.52d}$$

When x is equal to zero (at the boundary) and to integrals of one half-wavelength, $\cos kx = \pm 1$* so that $P = P_{max}$ and $\sin kx = \theta$, for which U is equal to U_{min},

$$p_{max} = p_0(1 + |r|) \quad \text{and} \quad U_{min} = u_0(1 - |r|). \tag{2.53a}$$

*It should be noted that the amplitude is always a positive quantity. The signs of p and u are taken care of by the combined effects of the signs of the functions of kx and ωt.

On the other hand, when x is equal to an odd number of quarter-wavelengths, $\cos kx = 0$ and $P = P_{min}$ and $\sin kx = \pm 1$, giving $U = U_{max}$,

$$p_{min} = p_0(1 - |r|) \quad \text{and} \quad U_{max} = u_0(1 + |r|). \qquad (2.53b)$$

Variations of amplitude, A, either P or U, with x are shown in Fig. 2.10.

2.6.3 R_2 is less than R_1

Here we have a 180° phase change for p but no phase change for u. This time r is negative in equation 2.51a and positive in equation 2.51b, so that:

$$p = 2|r|p_0 \sin kx \cos \omega t + (1 - |r|)p_0 \sin(\omega t - kx), \quad \text{and} \qquad (2.54a)$$

$$u = 2|r|u_0 \cos kx \sin \omega t + (1 - |r|)u_0 \sin(\omega t - kx). \qquad (2.54b)$$

Hence:

$$P = 2|r|p_0 \sin kx + (1 - |r|), \quad \text{and} \qquad (2.54c)$$

$$U = 2|r|u_0 \cos kx + (1 - |r|). \qquad (2.54d)$$

Hence, $P = P_{min}$ and $U = U_{max}$ at the boundary and at distances of a multiple of half-wavelengths from it; $\sin kx = 0$ and $\cos kx = \pm 1$, and with $P = P_{max}$ and $U = U_{min}$ at distances of odd numbers of quarter wavelengths from the boundary.

In both instances, when 100% reflection occurs, i.e. $|r| = 1$, it is seen that:

$$P_{max} = 2p_0, \quad P_{min} = 0, \quad U_{max} = 2u_0 \quad \text{and} \quad U_{min} = 0.$$

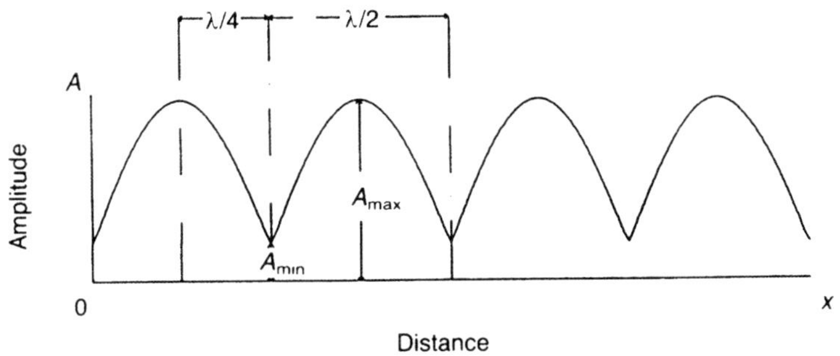

Fig. 2.10 Variation of amplitude, A, of either acoustic pressure or particle velocity with distance, x, from an arbitrary zero in a stationary wave system. λ denotes wavelength and the subscripts 'max' and 'min' indicate, respectively the maximum and minimum values.

The positions of maxima are called **nodes**, pressure or particle velocity nodes, and those of minima are termed **antinodes**. It can easily be shown that the particle displacement, y, behaves in the same way in stationary waves as the particle velocity.

2.6.4 Standing wave ratio

The ratio of the maximum to minimum amplitudes in a stationary wave system is called the **standing wave ratio (SWR)**:

$$\mathrm{SWR} = P_{max}/P_{min} = U_{max}/U_{min} = (1 + |r|)/(1 - |r|). \qquad (2.55)$$

This quantity varies from infinity, when there is 100% reflection, to unity for which there is no reflection and the waves are purely progressive.

2.6.5 Resonance of a stationary wave system

When stationary waves are formed in a uniform body terminated at each end by parallel boundaries perpendicular to the wave direction, the body as a whole can be considered as undergoing forced vibrations due to the action of the sound waves. There will be a number of resonance frequencies, as discussed below, and their values depend on the total length of the stationary wave system and on whether or not phase changes occur on reflections at the ends. The quality (Q) factors of the resonances depend on the attenuation coefficients of the material. Either half-wavelength or quarter-wavelength resonance occurs, depending on the boundary conditions at the ends.

Half-wavelength resonance takes place when the phase changes on reflection are the same at both ends of a body in which stationary waves are excited: either two nodes or two antinodes, so that its length is given by $l = n\lambda/2$ where n is an integer and indicates the order of the harmonic. The resonance or natural frequency at the nth harmonic is equal to $f_n = c/\lambda = nc/2l$. The first harmonic, f_1, is known as the fundamental frequency*. **Quarter-wavelength resonance** occurs when there is a node at one end and an antinode at the other so that:

$$1 = (2n - 1)\lambda/4 \quad \text{and} \quad f_n = (2n - 1)c/4\lambda,$$

and only odd harmonics are excited.

Resonance frequencies can be excited either singly, by applying sound waves at the required frequency by setting up forced vibrations, or in unison by exciting free vibrations. Using suitable restraints,

* This is at variance with the nomenclature used by musicians in which the first harmonic is the first overtone, which is our second harmonic and not the fundamental, and the terms overtone and harmonic are regarded as synonymous to them.

unwanted harmonics can be suppressed. For example, the even harmonics of a rectangular bar in air can be suppressed by clamping the latter at its mid-point at which a displacement antinode, i.e. a pressure node, would occur for these harmonics. Because of the effects of intertia, the amplitudes of the harmonics generally decrease as the order n increases.

2.7 DIFFRACTION

When sound waves encounter finite boundaries presented by apertures or obstructions, abrupt changes in directions of propagation may occur. This phenomenon is known as **diffraction**. An extreme case arising from obstructions having very small dimensions is that of scattering (section 2.4.3). Diffraction at a circular source and at sharp discontinuities, such as provided by crack tips, are dealt with in this section.

2.7.1 Diffraction from a circular source

Here we are concerned with diffraction associated with the finite boundary of a circular source of sound.

Consider a plane circular source of sound, located in an infinite baffle and vibrating with simple harmonic motion at a given frequency in a piston-like manner: the phase is uniform for all points on the vibrating surface. The effect of the infinite baffle is to ensure that radiation from only one of the surfaces is observed. We can then assume that the part of the source under consideration consists of a collection of a very large number of adjacent minute pulsating hemispheres, each emitting spherical waves having the same amplitude, phase and frequency which give rise to interference. It can be shown (for example, Kingsler *et al.* 1962) that the acoustic pressure amplitude, p_0, at an axial distance, r, in a homogeneous medium from the source of radius, a, is given by:

$$p_0 = |p_{max} \sin\{(k/2)[(r^2 + a^2)^{1/2} - r]\}| \tag{2.56}$$

where p_{max} represents the maximum value of acoustic pressure amplitude and k the wave number, $2\pi/\lambda$, λ being the wavelength.

The variation of p_0 with r along the axis of the source as given by the above equation is illustrated in Fig. 2.11(a), from which it can be seen that it is periodic, but with increasing distances between successive peaks. Values of p_0 vary from zero to a maximum p_{max}. p_0 reaches its penultimate trough where $r = a^2/2\lambda$ and its last maximum where $r = a^2/\lambda$. The final minimum value of p_0 occurs when r is infinite. Beyond the last maximum, where r becomes large compared with a,

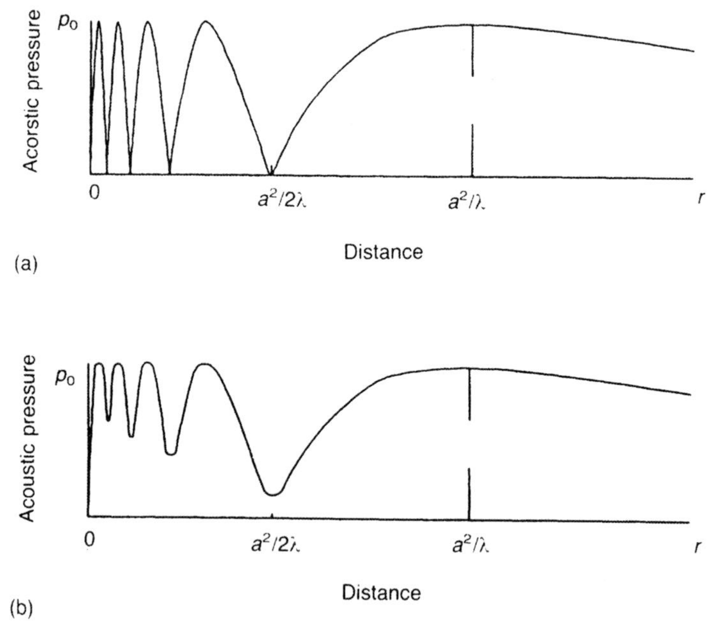

Fig. 2.11 Variation of acoustic pressure, p_0, with distance, r, along the axis of a circular source vibrating in a piston-like manner: (a) at a single frequency; and (b) over a wide frequency band (a = radius of the source and λ = wavelength (for centre frequency for (b)).

the sinusoidal term in equation 2.56 is sufficiently small for the equation to reduce to:

$$p_0 = p_{max} k a^2 / 4r^2 \tag{2.56a}$$

i.e. the amplitude decreases with $1/r$ as would occur when spherical waves radiate from a point at the centre of the source (sections 2.3.1, 2.3.5). It should be noted that equation 2.56 applies only to a single frequency: a frequency having an infinitesimal bandwidth. The region between the source and the last peak is known as the **near field** or **Fresnel zone**; the region beyond this peak is called the **far field** or **Fraunhofer zone**.

In practice, an ultrasonic source propagates over a finite bandwidth for which the curve represents the summation of the individual curves for every frequency within it. This results in a raising of the troughs and a broadening of the peaks (Fig. 2.11(b)).

It should be noted that equation 2.56 applies only to propagation along the axis of the source. For off-axis propagation, the equation becomes more complicated and the positions of the maxima and minima are displaced by amounts dependent on the distance of the beam from the axis. However, from considerations of conservation of energy, it is clear that the acoustic energy must remain constant for all cross-sections of the beam and a receiver of sufficiently large area of cross-section would indicate the same value of p_0 at all points within the near field; from a macroscopic point of view, plane wave conditions apply within the near field in a homogeneous medium.

In the far field, the spherical wave pattern may be complicated by the appearance of side lobes in addition to the main beam. At an off-axis point having polar coordinates (r, θ) with respect to the centre of the source, assuming axial symmetry, the acoustic pressure amplitude, p_0, is given by:

$$p_0 = |2p_{max}J_1(x)/x|, \tag{2.57}$$

where x is equal to the parameter $ka \sin\theta$. p_{max} is the value of p_0 on the axis $(r, 0)$ and $J_1(x)$ is a Bessel function of the first order (for example, Kreiszig, 1988) the value of which can be obtained from tables; k is equal to $2\pi/\lambda$ and a is the radius of the source.

Figure 2.12(a) shows (i) the variation of $2J_1(x)/x$ with x and (ii) the consequent variation of the ratio p_0/p_{max} with the parameter $\sin\theta(a/\lambda)$ (i.e. $ka \sin\theta/2\pi$) in accordance with equation 2.57. The first four zero values of p_0 are seen to occur when $\sin\theta$ is equal to $0.61\,\lambda/a$, $1.12\,\lambda/a$, $1.62\,\lambda/a$ and $2.12\,\lambda/a$. Figure 2.12(b) illustrates the variations of the ratio p_0/p_{max} with direction: directionality curves, when (i) $ka = \pi/2$, (ii) $ka = \pi$ and (iii) $ka = 5\pi$. With ka equal to π, a value for $\sin\theta$ of 1.22 is required to produce the first zero value of p_0, which is not possible because $\sin\theta$ must have a value of less than unity, and there is a considerable spread of the beam: i.e. the directivity is poor. However, when $ka = 5\pi$ as obtained, for example, by increasing either the frequency or the diameter of the source by a factor of five, the fourth zero value occurs when $\sin\theta$ is equal to 0.88 and the directivity curve indicates a main beam with three small lobes on each side of the axis. Most of the energy is in the main beam which can be seen to be highly directional; this is an essential feature with virtually all ultrasonic testing techniques.

Figure 2.13(a) illustrates the confines of the main beam propagated from a circular piston-like source. It is seen that the beam is parallel in the near field and, in the far field, it diverges so as to appear to originate from the centre of the source. This can be observed by means of a Schlieren technique (section 4.5.5). However, the usual and most convenient method of measuring acoustic pressure amplitudes is by observing echoes from small reflectors placed in the acoustic field as done by Krautkrämer with his distance gain size method (section 8.3.3). The effective lateral boundary of the beam is indicated by locating those points beyond which the echoes from the reflector are abruptly decreased in amplitude. In this way, the beam is effectively convergent in the near field and divergent, with a reduced angle of

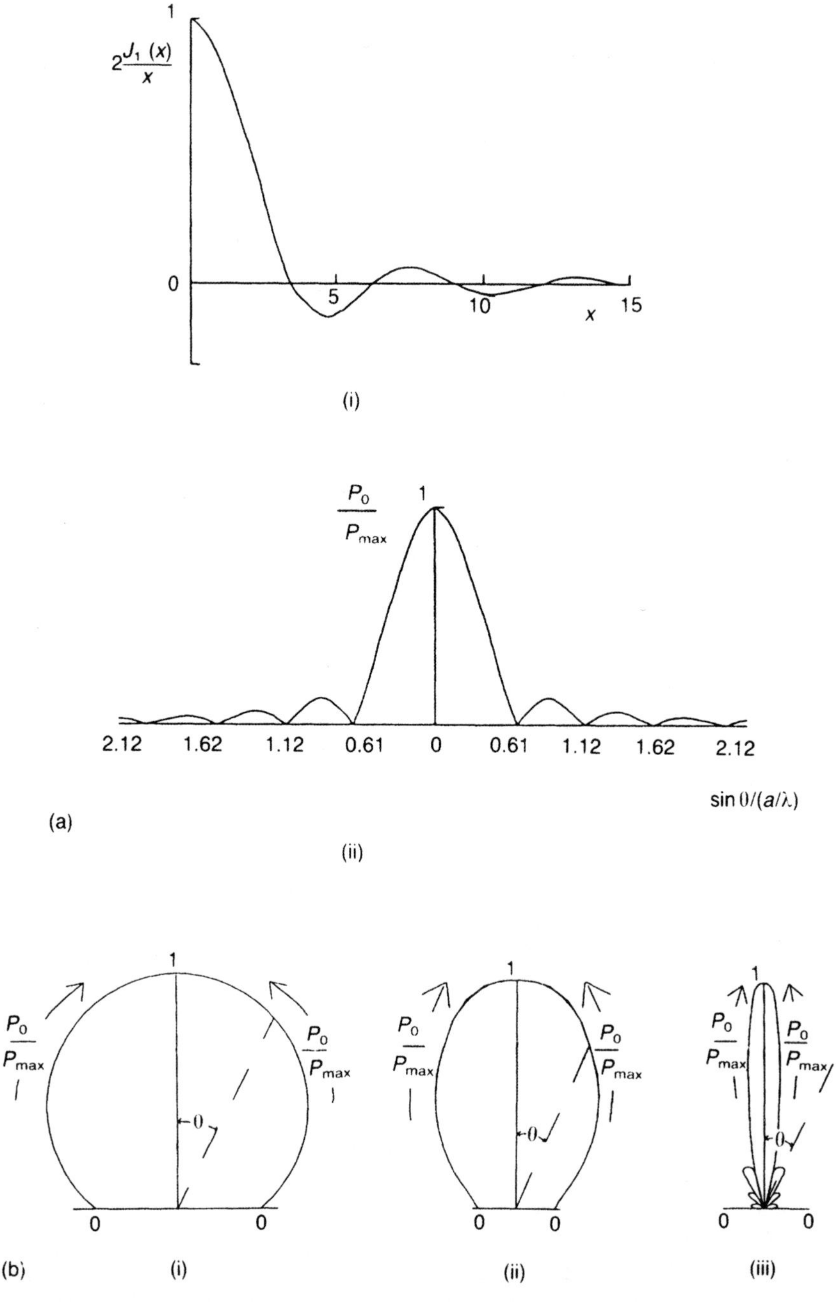

Fig. 2.12 (a) Variations of: (i) $2J_1(x)/x$ with x; and (ii) the consequent variation of the ratio p_0/p_{max} with the parameter $\sin\theta/(a/\lambda)$ (i.e. $ka\sin\theta/2\pi$) in accordance with equation 2.56. (b) Variations of the ratio p_0/p_{max} with direction, i.e. directionality curves when: (i) $ka = \pi/2$; (ii) $ka = \pi$; and (iii) $ka = 5\pi$.

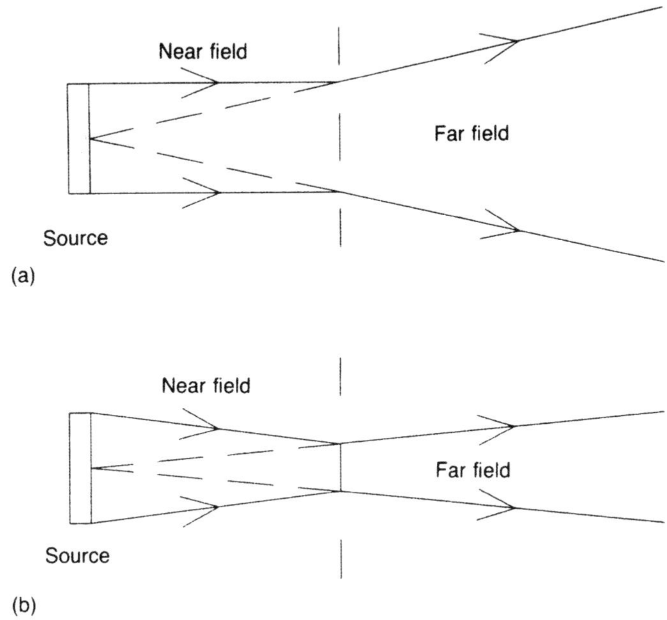

Fig. 2.13 Section though the main beam propagated from a piston-like source as obtained from: (a) theoretical considerations; and (b) measurements of echoes from small reflectors placed in the field. (Reproduced with permission from Krautkrämer 1990)

divergency, in the far field. Figure 2.13(b) indicates the beam pattern resulting from a decrease of 6 dB or more in the amplitude of the echo as the reflector is moved in a lateral direction. The lateral boundary of the beam in the near field is then approximately conical in shape.

2.7.2 Diffraction at sharp discontinuities

When ultrasound is incident at a linear discontinuity such as a crack, diffraction takes place at its extremities. The study of this phenomenon has led to the use of the time-of-flight diffraction method of sizing cracks. This has been used for many years for assessing the depths of surface cracks on concrete using low-frequency compression-wave probes with wide beam-spreads (section 10.2.7). A more recent development is the measurement of the sizes of non-surface-breaking cracks (section 8.2.3).

2.8 CONTINUOUS AND PULSED WAVES

A source undergoing forced vibrations (section 2.2.3) at a given amplitude and frequency for a time of duration, t, radiates continuous waves with a narrow frequency bandwidth centred at the exciting period. This bandwidth, δf, is equal to the reciprocal of t. For an extended time, e.g. 100 s, δf is equal to only 0.01 Hz: a very narrow frequency band. Thus $\delta f = 1$ Hz when $t = 1$ s and is equal to 10 Hz when $t = 0.1$ s. The amplitude remains constant throughout the entire duration except at the start and finish of the vibrations, which are characterized by the transients: the build-up and decay.

Pulsed waves can be generated either by interrupting continuous waves at regular intervals by a process known as gating, or by shock-exciting the source into free damped vibrations (section 2.2.2). The former method gives rise to long pulses for which the steady state predominates: e.g. for a frequency of 2 MHz and a pulse-width of 1 ms, the bandwidth is only 1 kHz. The latter method can provide very short pulses, e.g. 1 or 2 µs in width, corresponding to $\delta f = 1$ and 0.5 MHz, respectively.

2.9 PHASE AND GROUP VELOCITIES

The speed of sound, as given by equation 2.18, is the speed at which is given phase of vibration travels through a material and, for a specified direction, is known as the **phase velocity**. If the velocity is obtained by observing the time of travel acoustic energy as indicated by a position of maximum amplitude (e.g. the peak of a pulse), through a given thickness of a material, the **group velocity** is measured.

This phenomenon usually occurs with dispersive waves for which is speed of sound varies with frequency as experienced, for example, with the propagation of flexural waves in bars and with Lamb waves (section 3.5). A simple example is that of two waves having respective wavelengths λ and $\lambda + \delta\lambda$ with corresponding phase velocities v and $v + \delta v$, for which the group velocity v_g is given by:

$$v_g = v - \lambda\frac{\delta v}{\delta\lambda}. \tag{2.58}$$

With everyday ultrasonic testing, in which propagation is confined to bulk compression and shear waves at lower megahertz frequencies and at constant temperatures, velocity dispersion is virtually non-existent. However, with the broad frequency bands characteristic of pulsed waves, the group and phase velocities may not be the same and the difference between them increases with bandwidth (decreases with pulse-width). Consider, for example, a pulse of ultrasound passing through a polycrystalline metal sample in which attenuation is caused by Rayleigh scattering (section 3.3.1). The attenuation coefficient increases with the fourth power of the frequency and, consequently, the higher frequency contents of the pulse attenuate at a higher rate than the lower frequency contents as the energy progresses through the material.

To generate short pulses, the source, which is mounted for heavy damping, is shock-excited by means of a potential difference of several hundred volts across the electrodes of the transducer, thus providing an impulse pressure, p_0. The time of build-up of this potential difference is kept as short as possible, by proper circuit design, and is typically a quarter time-period, $T/4$, for free vibrations of the transducer. These vibrations are characterized by equation 2.7 with the variable p, the acoustic pressure, replacing y. Taking the exciting voltage, and hence p, to attain its maximum value when the time t is equal to $T/4$, the relationship between P and t can be expressed approximately (c.f. equation 2.7a) as:

$$p = p_0 \exp(\alpha' t) \sin \omega_0 t, \tag{2.59}$$

where p_0 is the amplitude of p, ω_0 the angular frequency $2\pi/T$ and α' the attenuation per unit time. Because of the combined effects of the finite time required for the voltage build-up and the damping of the transducer during this time, p cannot reach its peak value, p_0, as would be the case with undamped oscillations.

Figure 2.14 (a)(i) shows the variation of p with t for a transducer oscillating at a frequency of 4 MHz and a quality factor $Q = 2$ for α' equal to 6.3 neper μs^{-1}. This variation applies also to the layer of the material of propagation immediately in contact with the transducer surface. Figure 2.14(b)(i) illustrates the corresponding frequency distribution curve for which the maximum, for which $\omega = \omega_0$, lies at a frequency of 4 MHz. This curve is obtained by performing a Fourier transform (for example, Champeney 1973) using the following relationships:

$$f(t) = (1/2\pi) \int_{-\infty}^{\infty} f(\omega) \exp(j\omega t) d\omega, \tag{2.60a}$$

where $f(t)$ is a function of time, y in equation 2.59, and $f(\omega)$ a function of frequency, and

$$f(\omega) = \int_{-\infty}^{\infty} f(t) \exp(j\omega t) dt. \tag{2.60b}$$

Equation 2.60b is the Fourier transform of equation 2.60a and equation 2.60a the inverse transform of equation 2.60b. It can thus be shown that the Fourier transform of equation 2.59 is:

$$P = 2p_0\alpha'/[\alpha'^2 + (\omega_0 - \omega)^2], \tag{2.61}$$

where P is the acoustic pressure amplitude for a given angular frequency ω.

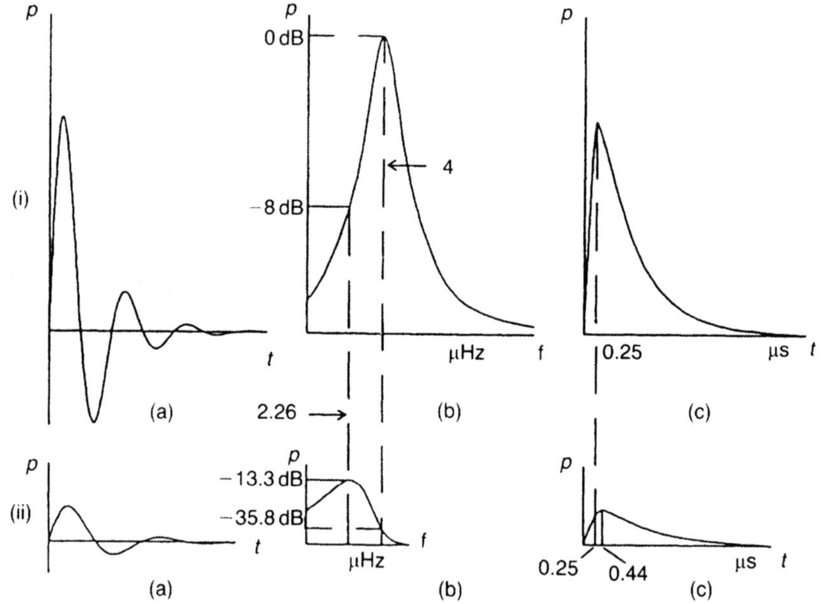

Fig. 2.14 (a) Variation of acoustic pressure, p, with time, t, in an ultrasonic pulse resulting from the free damped vibrations of an acoustic source subjected to an initial pressure, p_0, in accordance with equation 2.58: (i) on propagation; and (ii) after attenuation. (b) Variations of acoustic pressure amplitudes with frequency in the pulses as obtained by the Fourier transforms of the curves in (a). (c) Curves showing the corresponding rectified and smoothed pulses.

As indicated in section 2.3.1, the vibrations of the source are transmitted through the material under test and the amplitude of vibration of an imaginary layer of the material at some distance x from the source depends on the attenuation coefficient of the waves and on any divergence of the beam, as in the far field. The maximum amplitudes of the peaks contained by the pulse decrease as x increases as a result of attenuation. When the attenuation is frequency dependent as, for example, with Rayleigh scattering (section 2.4.3), the corresponding frequency distribution curve for the pulse, which depends on the value of x, changes its shape. Figure 2.14(b)(ii) illustrates this type of curve for a received pulse after an attenuation of 35.8 dB for the original centre frequency of 4 MHz. The peak now lies at 2.26 MHz. In the original curve (Fig. 2.14(b)(i)) the value of P corresponding to this frequency is 8 dB below that of that of P_0, for 4 MHz. The resultant attenuation corresponding to 2.26 MHz is only 5.3 dB, as predicted by equation 3.9. The peak of the pulse is thus attenuated by 13.3 dB (8 + 5.3).

Figure 2.14(a)(ii) shows the inverse transform of (b)(ii) from which the right-hand shift of the first peak indicating the increase in time period can be clearly seen. (c)(i) and (c)(ii) illustrate the rectified and smoothed curves corresponding to those in (a)(i) and (a)(ii), respectively, as would normally appear on the screen of an ultrasonic flaw detector. They clearly show the shift in position of the peak of the pulse. For propagation in steel having a speed of sound of $6000\,m\,s^{-1}$ the indicated additional time delay of $0.19\,\mu s$ of the peak corresponds to a shift of 1.14 mm.

It should be noted that, for the sake of simplicity, the occurrence of the 'dead zone' arising from overloading the transducer has been ignored (section 5.3.1). For comparing pulses, this difficulty can be overcome either by using a buffer and comparing front- and back-wall echoes or by comparing different back-wall echoes.

Measurement of the time of transit of the peak of the pulse for a given path length produces a value of the **group velocity** of the energy, this is clearly less than the corresponding **phase velocity**. However, the latter can be estimated with sufficient accuracy for most purposes by noting the position of the leading edge of the pulse on the time scale instead of the peak, as clearly illustrated by Figs 2.14(c)(i) and (c)(ii). Attenuation measurements at given frequencies and also more precise time-delay measurements can be obtained by using a narrow-band filter tuned to the required frequency.

2.10 FOCUSING SYSTEMS

With B- and C-scan and other ultrasonic imaging immersion techniques (sections 5.6 and 7.5), an improved degree of image resolution can be obtained by focusing the sound waves by means of a converging lens in front of a transducer with a concave radiating surface or an array of transducers mounted on a concave surface.

An acoustic lens is designed in a similar fashion as an optical lens (for example, Jenkins and White, 1976) and its focal length f_L is given by the following expression:

$$1/f_L = |(c - c_L)/c_L|(1/r_1 + 1/r_2), \quad (2.62)$$

where r_1 and r_2 represent the radii of curvature of the lens surfaces, and c and c_L the speeds of sound in the surrounding medium and the lens, respectively. Postulating $|(c - c_L)/c_L|$ as a positive quantity, the appropriate radius of curvature is positive for a convex surface of the lens in contact with the liquid when c is greater than c_L; this is equivalent to the real-is-positive convention used in optics. For optimum coupling, the characteristic impedances of the lens and medium should be of the same order of magnitude. Considering, for example, a Perspex lens in water, the data in Table 2.1 indicate that the transmission coefficient for each surface is about 75% if we conveniently disregard the fact that the lens is a not a semi-infinite medium. Using a **planoconvex** system where the radius of the curved surface is 100 mm, a **converging lens** having a focal length of 170 mm is obtained. This is in contrast with the effect of a glass lens in air for which a planoconcave system is divergent, because the speed of sound in an acoustic lens is usually greater than that in the surrounding medium whereas the speed of light in an optical lens is less than that in air. Figure 2.15(a) illustrates the formation of an image from an object irradiated by a transducer.

Equation 2.62 is valid only for paraxial beams, as with optics. This condition is best achieved by reducing the amout of deviation by using a planoconcave lens of low curvature either singly or in tandem, as with optical instruments. It is important that the frequency bandwidth be kept as narrow as possible to avoid the blurring of images as found with chromatic aberration experienced with optical instruments.

With a parallel beam incident at the lens, the beam converges at the principles focus. With this arrangement, it may be convenient for the plane surface of a planoconcave ultrasonic lens to be permanently bonded to the transducer surface. If the focal length is made equal to the length of the near field the spread of the beam in the far field is eliminated (Fig. 2.15(b)). It is sometimes advantageous to use a ceramic transducer having a ground concave surface. Devices having these arrangements are called **focused probes**.

Focused probes, like all lens systems, are susceptible to chromatic aberration. For many imaging applications this may not be a problem but, for high resolution imaging the use of the more sophisticated phased-array transducers eliminates these defects (section 4.4.4).

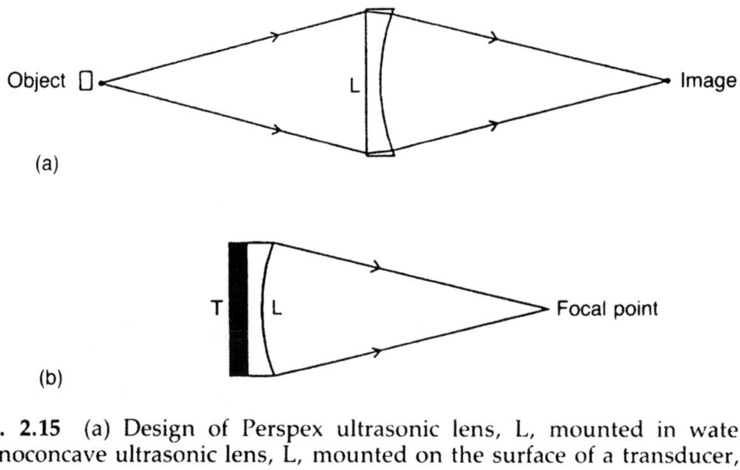

Fig. 2.15 (a) Design of Perspex ultrasonic lens, L, mounted in water. (b) planoconcave ultrasonic lens, L, mounted on the surface of a transducer, T.

REFERENCES

Champeney, D.C. (1973) *Fourier Transforms and their Physical Applications*, Academic Press, London.

Jenkins, F.A. and White, H.E. (1976) *Fundamentals of Optics*, 4th edn, (International Students edn). McGraw-Hill: Tokyo.

Kinsler, L.E., Frey, A.R., Coppens, A.B. and Sanders, J.V. (1982) *Fundamentals of Acoustics*, 3rd edn. Wiley, New York.

Kreiszig, K. (1988) *Advanced Engineering Mathematics*, 6th edn. Wiley, New York.

Mayer, W.G. (1965) *Ultrasonics*, **3**, 63–68.

CHAPTER 3

Ultrasonic characterization

3.1 GENERAL CONSIDERATIONS

The value of the speed of sound in a material depends on the appropriate elastic modulus, q, which is characteristic of the nature of the vibrational stresses applied to it and also on the density of the material, as shown in section 2.3.3 (equation 2.18). It is assumed that the amplitudes of the stresses are sufficiently low to allow for purely elastic conditions and that q represents the adiabatic value q_S of the elastic modulus, which indicates that the variations of the stresses and strains in acoustic waves occur so rapidly that there is no time for heat exchanges to occur. This is in contrast with the isothermal value, q_T, evaluated under static conditions when enough time is available for thermal equilibrium to be established. The relationship between these two moduli is given by:

$$\gamma = q_S/q_T, \tag{3.1}$$

where γ is the ratio of the principal specific heats, at constant pressure and constant volume for a given temperature. This is approximately equal to 1, to within 1%, for most solids and liquids but may have a value of up to 1.67 for gases.

3.2 ULTRASONIC SPEEDS

The appropriate elastic modulus governing the value of the speed of sound depends on the type of waves which are propagated and the relative values of the wavelength and cross-sectional dimensions of the irradiated object. For compression waves propagated along the axial direction in a bar having lateral dimensions small compared with the wavelength, $q = E$, the Young modulus, which is the ratio of the applied tensile stress to the resultant strain in the direction of the stress. The speed is given by:

$$c = (E/\rho)^{1/2}, \tag{3.2}$$

where ρ is the density (for solid rods). Because there are no lateral constraints imposed by the material of the rod, compressions and tensions in the axial direction give rise to expansions and contractions, respectively, in the plane of the cross-section by amounts determined by the **Poisson ratio**, ν, as defined by:

$$\nu = -\varepsilon_2/\varepsilon_1, \tag{3.3}$$

where ε_1 and ε_2 are respectively the axial and linear cross-sectional strains: the negative sign takes care of the fact that, for a positive axial strain, the lateral strain is negative. With isotropic materials, ε_2 is constant in all directions perpendicular to the axis. It is easy to visualize that when compressive (longitudinal) waves travel along the axis of the rod, transverse surface waves, having the same speed as the compressive waves, appear. These should not be confused with the shear waves in bulk solids or with Rayleigh surface waves (section 3.5).

Ultrasonic testing is carried out mainly at megahertz frequencies where the wavelength is generally small compared with the cross-sectional dimensions of the object being tested. For example, let us suppose that compression waves of frequency 2 MHz are transmitted through a steel block having a 50 mm square cross-section. At this frequency, the wavelength is of the order of 3 mm. The lateral strains arising from the longitudinal stresses are opposed by the inertia of the bulk of the material and a more complex situation arises because the relationship between tensile stress and tensile strain depends on the Poisson ratio, ν, as well as the Young modulus, E. The appropriate elastic constant is the **axial modulus**, χ, which can be shown (for example, Sprackling 1970) to be given by:

$$\chi = K + 4G/3 \tag{3.4a}$$

By making the substitutions $K = E/3(1 - \nu)$ and $G = E/2(1 + \nu)$, it can be seen that an alternative expression for χ is given by:

$$\chi = E(1 - \nu)/[(1 + \nu)(1 - 2\nu)], \tag{3.4b}$$

where K is the bulk modulus (the ratio of the bulk stress to the volume strain) and G the shear modulus, or rigidity (the ratio of the shear stress to the shear strain) of the material.

Thus the speed of compression waves is then given by either

$$c = [(K + 4G/3)/\rho]^{1/2} \quad \text{(bulk solids)} \tag{3.5a}$$

or

$$c = \{E(1 - \nu)/[\rho(1 + \nu)(1 - 2\nu)]\}^{1/2} \quad \text{(bulk solids).} \tag{3.5b}$$

The first of these expressions appears to be favoured mainly by physicists and the second by engineers.

Liquids and gases cannot support shear stresses, and the value of the shear modulus for these materials is zero. Equation 3.5 then becomes:

$$c = (K/\rho)^{1/2} \quad \text{(fluids).} \tag{3.6a}$$

The speed of sound in fluids is sometimes expressed as:

$$c = [(1/(\beta\rho)]^{1/2}, \qquad (3.6b)$$

where $\beta = 1/K$ is the **compressibility**. For an 'ideal gas', the speed of sound can be shown to be:

$$c = (\gamma P/\rho)^{1/2} \quad \text{(ideal gases)}, \qquad (3.7)$$

where γ is defined above, and P the ambient pressure. The speed of shear waves is given by:

$$c = (G/\rho)^{1/2} \quad \text{(shear waves)}. \qquad (3.8)$$

Equation 3.8 is independent of the dimensions of the material.

Table 3.1 enumerates some values of acoustic speeds in solids in both bar and bulk forms and in fluids. These values should be regarded as typical rather than exact because of their dependence on a number of factors governing the values of the elastic constants, as indicated below. In most instances the speed of shear waves in a given bulk solid is of the order of one half that of compression waves, with lead and

Table 3.1 Approximate values of speeds of compression and transverse waves in some commonly used materials at room temperature. The data given comprise typical values. Exact values depend on structure, previous history, impurity content and physical conditions

Material	*Speed* ($m\,s^{-1}$)		
	Compression waves		*Shear waves*
	Bulk	*Rods*	
Aluminium	6400	5100	3100
Brass	4400	3500	2100
Copper	4800	3700	2300
Gold	3200	2000	1200
Lead	2200	1200	700
Mild steel	6000	4800	2900
Nickel	5600	4800	3000
Platinum	3300	2800	1700
Silver	3700	2800	1700
Tin	3400	2600	1600
Nylon	2600	1800	1100
Perspex	2700	2100	1300
Silica (fused)	6000	5800	3800
Lead zirconate titanate (PZT)	4800		
Glycerol	1900		
Lubricating oil	1400–1800		
Water	1500		
Air	330		

gold providing notable exceptions. Because the elastic moduli and densities of substances vary with temperature, the speeds of sound also do so. With rise in temperature, there is generally a decrease in speed for solids and liquids and an increase for gases. For water, however, there is an increase in speed with rise in temperature to a maximum of 73°C at normal pressure, and then a decrease with further temperature rise.

If the speed of sound in a parallel-sided object is known, the perpendicular distance between the opposite sides can be determined by measuring the time taken for the waves to travel from one to the other. Conversely, if the dimensions and density are known, the elastic modulus appropriate to the nature of the sound waves can be evaluated. Elastic moduli vary in magnitude with temperature and physical properties of the material; e.g. in metals, with degree of hardness, grain size, impurity content, the relative proportions of the constituents of an alloy and the values of any applied load and internal stresses. For example, internal stresses in a sample undergoing a fatigue test can be monitored by measuring the speed of sound in it.

3.3 ULTRASONIC SCATTERING AND ABSORPTION

3.3.1 Scattering

Virtually all metals used in the manufacture of structures and components are polycrystalline in form: they are made up from a large number of small randomly orientated individual grains which are formed when the metal is originally cast. Each of these grains is a single crystal and, in general, has anisotropic properties (section 3.4). A single crystal is usually characterized by its axes, usually orientated in three mutually perpendicular directions, along which the various physical properties, such as elastic moduli, have unique values. In polycrystalline metals, the axes within the individual grains are orientated at random with respect to those in the neighbouring grains. However, because there is a very large number of these grains, the overall effect is that the anisotropy disappears and the material as a whole is effectively isotropic provided that there are no preferred orientations brought about by external stresses, as in rolled steels.

When sound travels in a given direction through a polycrystalline substance there is a discontinuity in its speed at each grain boundary, because of changes in the elastic moduli and, hence, characteristic impedances (section 2.5) as a result of the corresponding changes in the directions of the axes of the grains. Reflections thus take place at the grain boundaries and scattering takes place when the mean linear dimension, D, of the grains is comparable with, or smaller than, the wavelength. The scattered energy is then lost from the beam thus providing attenuation. Special circumstances arise when the dimensions of the reflecting surfaces are very small compared with the wavelength λ, as when $\lambda \geqslant D/10$, in solid suspensions in liquids and porosity in solids. Here, the scattering takes place equally in all directions (Rayleigh scattering) and the attenuation coefficient, α, is given by:

$$\alpha = kD^3/\lambda^4 \tag{3.9a}$$

$$\alpha = KD^3f^4 \tag{3.9b}$$

where λ is the wavelength and f the frequency of the waves. k is a constant having a value which depends on the degree of anisotropy of the material. $K = k/c^4$ is a constant for any given speed of sound and, provided that concern is only with a given material at a fixed temperature, one can use the more convenient variable of frequency. With increasing wavelengths (decreasing frequency) the conditions for Rayleigh scattering may no longer exist and there is a reduction in attenuation due to this phenomenon (Huntingdon 1940). Rayleigh scattering is not usually significant at frequencies of less than 1 MHz unless caused by distributed porosity and inclusions having dimensions larger than those of the grains which are present.

Rayleigh scattering is a highly significant factor for considering the choice of frequency for testing. Equation 3.9 shows that if, for a given metal, the attenuation over a distance of 100 mm is only 1 dB at a frequency of 1 MHz, it increases to 625 dB at a frequency of 5 MHz due to this phenomenon because of the fourth power dependence on frequency, assuming that the conditions for Rayleigh scattering remain the same. The dependence on grain dimension enables the estimation of grain size from attenuation measurements. In section 2.9 it is shown how the attenuation caused by Rayleigh scattering causes changes in shape of ultrasonic pulses (Fig. 2.14) as a result of their broad-band frequency content.

In concrete, scattering arises from the effects of the aggregate in its structure, which may produce values of D of up to 20 mm and, considering a typical acoustic speed of 4000 m s^{-1}, Rayleigh scattering can be observed with frequencies in the lower kilohertz range thus setting an upper frequency limit of 20–100 kHz, depending on the aggregate size, for testing this material.

Mention should be made of the phenomenon of **backscatter**: that part of the scatter taking place in the direction of the acoustic source. This phenomenon can allow observations of surface roughness, sub-surface microstructure and hardening depth as well as monitoring of stress. Because the degree of scatter increases with frequency, greater sensitivity of measurement is possible at higher frequencies. Measurements of backscatter are discussed briefly in section 5.8.

3.3.2 Absorption in solids

There are a number of causes of absorption of ultrasound in solids but, for polycrystalline metals tested at megahertz frequencies, the resulting attenuations are usually negligible compared with those caused by Rayleigh scattering. One cause of absorption in polycrystalline metals is friction at grain boundaries for which the corresponding attenuation coefficient increases in direct proportion to the frequency. This can be detected only at frequencies sufficiently low for the masking effect of scattering to be negligible.

Ultrasonic absorption may arise from the presence of microscopic structural defects which can occur in both metals and non-metals and consist of two kinds: point defects and dislocations. The presence of these defects can affect the degree of hardness and also the elastic properties of the material. Dislocations may be described as discontinuities in the lattice structure of a crystalline solid. Point defects comprise lattice vacancies (atoms missing from the lattice structure), interstitial atoms (atoms not occupying a proper place in the lattice structure) and impurity atoms, either interstitial or substitutional (substituting a 'native' atom in the lattice structure). Any of these defects may take up energy from the sound passing through the material and thus give rise to absorption. The corresponding attenuation coefficient depends on both frequency and temperature.

These structural defects are considerably smaller than macroscopic defects. In particular, dislocations can be set into motion and multiply in number either as a result of direct stressing or fatigue. This process may eventually lead to the appearance of cracks. The behaviour of a fatigue process can be monitored by measuring the ultrasonic attenuation in the materials, preferably at the frequency at which the absorption is maximum. Because the effects of dislocations on sound waves are significant at frequencies of 10–50 kHz, the damping capacity method (section 5.7.2) is often used for these measurements.

Another type of absorption occurs with ferromagnetic metals such as nickel, cobalt and most steels and also with ferrimagnetic substances called ferrites. These latter materials are strongly ferromagnetic and sound energy passing through them may be partly coverted into magnetic energy during the positive half-cycle and back to sound energy during the negative half-cycle. The phenomenon, which is somewhat complicated, is discussed in some detail by Mason (1959). As a result of phase changes occuring during the energy exchanges, hysteresis occurs between the acoustic pressure and particle velocity so that the resulting attenuation coefficient is dependent on both amplitude and frequency. The loss per cycle can be evaluated from the area of the acoustic pressure against particle velocity hysteresis curve. The attenuation generally reaches a maximum at kilohertz frequencies and becomes comparatively low at megahertz frequencies. A similar phenomenon occurs with ferroelectric materials, including lead metaniobate, PZT, barium titanate and other substances used for ceramic ultrasonic transducers (section 4.2). This accounts for their low '*Q*' factors and the consequent high temperature rises when they are excited continuously.

A fuller account of absorption in solids is given by Beyer and Letcher (1969).

3.3.3 Absorption in fluids

It is stated in section 3.2 that liquids and gases cannot support shear stresses: a fluid collapses very rapidly when subjected to a shear stress. It is also stated that the axial modulus associated with bulk compression waves has a shear component. The collapse of the fluid gives rise to frictional losses and absorption consequently takes place. This phenomenon is called **shear viscosity** and the corresponding attenuation coefficient is dependent on the coefficient of viscosity of the fluid, as defined by Newton's law of viscosity: the shear stress is proportional to the **rate of change** of strain, assuming streamline conditions (the absence of turbulence).

In certain liquids, including water and alcohol, known as associated liquids, local collapses of the lattice structure occur as a result of an applied pressure. Collapse occurs during the positive half-cycle of compression waves and recovery takes place during the negative half-cycle. The consequent friction gives rise to absorption and the phenomenon is known as **bulk viscosity** or structural relaxation. Other possible causes of absorption in liquids include thermal conduction and thermal radiation (Beyer and Letcher 1969). It should be noted that two or more absorption mechanisms can act simultaneously and the corresponding attenuation coefficients are additive.

3.4 ANISOTROPIC MATERIALS

The values of the speeds of sound quoted in Table 3.1 apply only to isotropic samples of the materials concerned, where the physical properties remain the same for all directions. Most materials used for the manufacture of engineering components, such as polycrystalline metals, are isotropic but there are a number for which some of the physical properties, elastic moduli, electrical resistance and thermal conductivity, vary with direction. These latter materials are called **anisotropic** and include most single crystals, materials subjected to uniaxial stresses and materials given preferential treatment in a particular direction, such as rolled steel sheet. The speed of sound in anisotropic materials varies with direction and its value for shear waves also depends on the direction of the particle vibrations (the polarization of the waves). Measuring speeds of sound for various directions and polarizations (for shear waves) enables one to determine the degree of anisotropy in a solid.

For example, with single crystal germanium, for propagation in the [100] direction, the speed of compression waves is $4900\,\mathrm{m\,s^{-1}}$ and that for shear waves is $3500\,\mathrm{m\,s^{-1}}$, for all directions of polarization. In the [110] direction, however, the speed of compression waves is $5400\,\mathrm{m\,s^{-1}}$ and, for shear waves, the speed is $3500\,\mathrm{m\,s^{-1}}$ for polarization in the [010] direction and $2700\,\mathrm{m\,s^{-1}}$ for polarization in the [110] direction. These values reflect the anisotropy of the corresponding elastic constants. Relationships between the acoustic speed and elastic constants for anisotropic materials are discussed by Beyer and Letcher (1969).

3.5 SURFACE AND LAMB WAVES

It is shown in section 2.5.2 how **Rayleigh** surface waves can be propagated across the surface of a solid. With surface waves, the vibrations are confined to the neighbourhood of the surface and the acoustic energy penetrates the solid by a distance of not greater than one or two wavelengths; for this reason the attenuation of surface waves is considerably less than that for bulk waves. The polarization of Rayleigh waves lies in a plane perpendicular to the surface, as discussed below. Another type of surface waves is **Love waves** for which the polarization lies in the plane of the surface but perpendicular to the direction of propagation. However, they have an upper frequency limit of a few thousand hertz and are normally encountered only with seismic applications.

The vibrations of the particles for Rayleigh waves have two components, perpendicular to and parallel with the surface, as for longitudinal (compression) and transverse (shear) waves, respectively. Because the components travel at different speeds, the vibrations are polarized elliptically with a continual change in phase and the shape of the ellipse varies in a cyclical manner (Fig. 3.1). The speed, c_R, of Rayleigh waves is given by a complicated expression (Redwood 1960), but it can be expressed approximately as:

$$c_R = 0.9c_T, \tag{3.10}$$

where c_T is the speed of shear waves for the bulk materials.

For an object sufficiently thin to allow penetration to the opposite surface, e.g. a plate having a thickness of the order of a wavelength or so, Rayleigh waves degenerate to **plate** or **Lamb** waves, whch can be propagated in a number of modes, either **symmetrical** or **anti-**

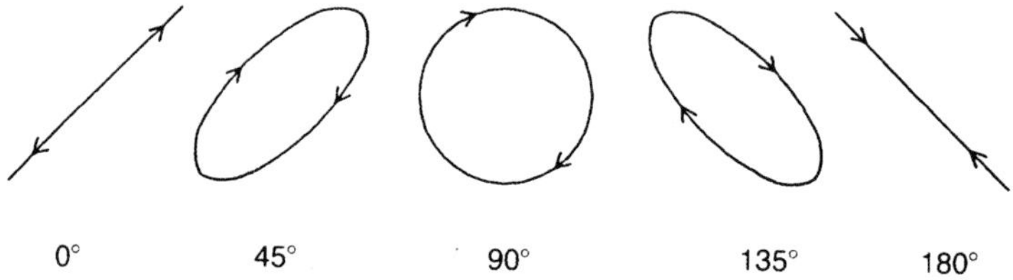

Fig. 3.1 Variation of elliptical polarization of Rayleigh waves with phase angle difference between the longitudinal and transverse components. The ellipses for 225° and 315° are the respective mirror images of those for 135° and 45°.

symmetrical (Figs 3.2 and 3.3). The speed of Lamb waves varies with frequency and, for this reason, a distinction should be made between phase and group velocity (section 2.9). The value of the phase velocity for each mode depends on the order of the mode: fundamental (0th), 1st, 2nd, 3rd, etc., whether the mode is either symmetrical or antisymmetrical, the product of the frequency, f, and plate-thickness, d, and the nature of the material of the plate. The relations between these quantities are highly complicated and the interested reader is referred to Worlton (1957) for further information.

It has been shown, for example, that the phase velocity for the first antisymmetrical mode is approximately $2600\,\mathrm{m\,s^{-1}}$ in aluminium for all values of the product, fd, greater than about 5 MHz mm and, for a frequency of 5 MHz, the corresponding value of d is approximately 1 mm. Lamb waves can be used for detecting laminar defects of up to this order of distance below the surface, because the material between the surface and the lamination itself forms a plate. Applications of Lamb waves to ultrasonic testing are discussed in sections 5.10 and 10.7.4.

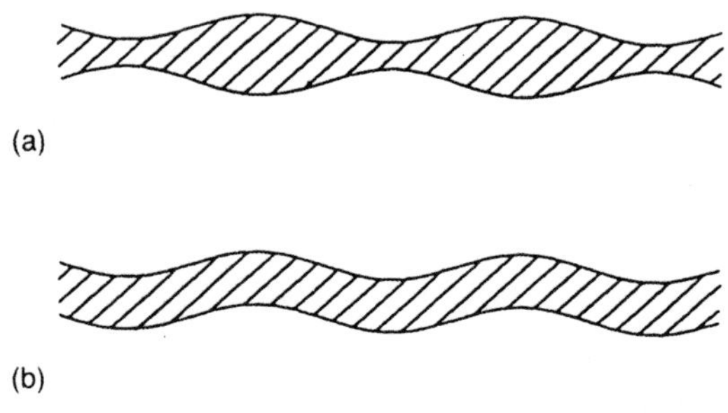

Fig. 3.2 Configurations, at a given time, of the cross-section of a plate carrying Lamb waves for: (a) symmetrical; and (b) antisymmetrical modes. The displacements in the vertical directions are grossly exaggerated for clarification.

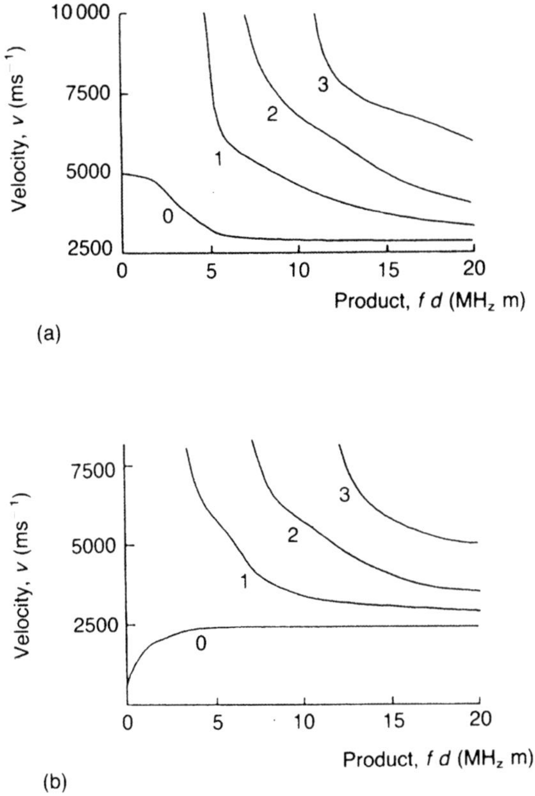

Fig. 3.3 Relationships between the phase velocities, v, and the product of frequency, f, and plate-thickness, d, for: (a) symmetrical; and (b) antisymmetrical Lamb waves. Numerals indicate orders of modes (after Worlton (1957)).

REFERENCES

Beyer, T.R. and Letcher, S.V. (1969) *Physical Ultrasonics*. Academic Press, New York and London.

Huntingdon, H.B. (1940) *J Acoust Soc Am*, **22**, 362–4.

Mason, W.P. (1959) *Physical Acoustics and the Properties of Solids*. van Nostrand, Princeton.

Redwood, M. (1960) *Mechanical Waveguides*. Pergamon, Oxford, Section 2.4, pp. 50–51.

Sprackling, M.T. (1970) *The Mechanical Properties of Solids*. EUP, London, p. 29.

Worlton, D.C. (1957) *Nondestr Test*, **15**, 218–22.

CHAPTER 4

Ultrasonic transducers

4.1 INTRODUCTION

A transducer is a device that converts energy from one form to another, e.g. electrical energy to ultrasonic energy or vice versa. Most types of ultrasonic transducers used for non-destructive testing are reversible: the energy conversion is performed in both directions with equal efficiencies.

Ultrasonic transducers may be classified as follows.

- **Piezoelectric**: reversible and most widely used for NDT but normally requiring contact with the test sample.
- **Electromagnetic**: reversible and can be used for testing metal samples without making contact with them.
- **Electrostatic**: reversible and used mainly for low-frequency applications such as damping capacity measurements. They should be located as closely as possible to the test sample but without actually touching it.
- **Magnetostrictive**: reversible and also mainly restricted to low frequency applications. Except for the testing of ferromagnetic materials, which effectively form part of the transducer arrangement, contact with the test sample is necessary.
- **Laser and other optical**: not reversible. The energy for laser beams can be directly converted into acoustic energy and, for this reason, a laser device can be used to transmit ultrasound. Both laser beams and conventional optical beams can be used as receivers. No material contact is involved and, unlike other ultrasonic devices, laser and other optical transducers can be located remotely from the object under test.
- **Miscellaneous**: including devices making use of mechanical, thermal and chemical energy.

Important considerations in the choice of transducer are quality of performance, ease of application of cost. For the vast majority of uses, the piezoelectric transducer is most suitable in these respects. However, for transmission, the ultrasonic waves have to be propagated in some way through the object under test and this requires the use of a suitable coupling material to provide a satisfactory transmission coefficient (section 2.5). The same problem arises for a receiving transducer. This may not be possible for testing in a hostile environment, such as at high temperatures, or when physical contact of a liquid couplant with the material is undesirable as, for example, with a hygroscopic solid. Some of the other types of transducer listed above may prove to be more suitable because the ultrasound is induced directly into the material under test. A rare exception, of course, would arise with the testing of a piezoelectric object. The coupling of piezoelectric transducers is discussed in section 4.4.2.

4.2 PIEZOELECTRIC TRANSDUCERS: PRINCIPLES

4.2.1 The piezoelectric effect

The **piezoelectric effect** is characteristic of single crystals of certain insulators, including quartz and lithium sulphate, and some semiconductors such as cadmium sulphide and zinc oxide. The effect is also observed with suitably prepared ceramic materials such as barium titanate, lead zirconate titanate (PZT) and lead meta-niobate, which possess strong electrostrictive properties and are normally used in ultrasonic flaw detection probes.

The piezoelectric effect requires the presence of a polar axis. This axis is present naturally in the single crystal materials used as transducers but it must be induced artificially in the electrostrictive ceramics (section 4.3.2). Consider a disc made from one of these materials having a polar axis perpendicular to the flat parallel surfaces. On subjecting it to a mechanical stress, equal and opposite electric charges appear on these surfaces. This is the **direct piezoelectric effect**. Figure 4.1(a) illustrates the phenomenon for a tensile stress acting along a polar axis characteristic of longitudinal stresses and strains, e.g. the x-axis of a quartz crystal or the z-axis of a ceramic transducer. On replacing the tensile stress by a compressive one of equal amplitude, the surface charges exchange polarity but do not change in strength (Fig. 4.1(b)). Provided that the elastic limit is not exceeded as a result of the stressing, the polarization (the charge per unit area), P, at the surfaces is proportional to the applied stress σ:

$$P = \sigma d \text{ (short-circuit).} \tag{4.1}$$

P is expressed in coulombs per square metre ($\mathrm{C\,m^{-2}}$) and σ in pascals (Pa, $\mathrm{N\,m^{-2}}$). Polarization is observed in addition to the mechanical strain produced by the stress. In practice, the surfaces of the plates are thinly coated with metal electrodes and P can be evaluated by

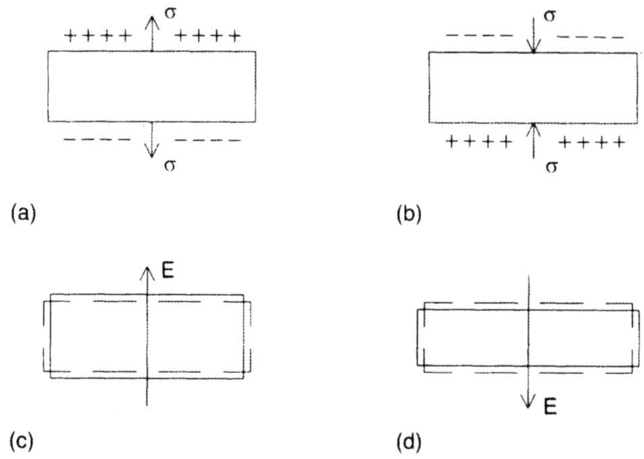

Fig. 4.1 Compressions and expansions of a piezoelectric transducer: (a) effect of tensile stress σ; (b) effect of compressive stress σ; (c) expansion caused by application of electric field E; and (d) contraction caused by reversal of E. The broken lines in (c) and (d) indicate the dimensions of the transducer when not in operation.

measuring the flow of charge when the electrodes are short-circuited; d is a constant for a given direction of polar axis and is called the **piezoelectric strain constant** or simply the 'd' coefficient and expressed in coulombs per newton ($C N^{-1}$).

The example illustrated by Figs 4.1 (a) and (b) is valid only when the direction of the polar axis is suitable for the productions of tensile stresses and strains characteristic of compression waves. For the direct production of shear stresses and strains, the polar axis (e.g. the y-axis of a quartz crystal) should be perpendicular to the thickness direction of the transducer, but this would give rise to difficulties for propagation into solids when using liquid couplants, which collapse under the action of shear stresses. The usual method of transmitting shear waves through solids is the use of a mode-conversion technique (section 2.5.2).

With the **converse piezoelectric effect** an electric field, E, applied along the polar axis gives rises to a mechanical strain, S*, and, provided that there is no mechanical loading on the disc, S is directly proportional to E, assuming elastic conditions:

$$S = Ed \text{ (no load).} \tag{4.2}$$

This phenomenon occurs in addition to the production of electric charges by the field. Because the effect is reversible, the constant d is

*The mechanical strain is usually identified by the Greek letter ε. Here the letter S is used in place of ε to avoid confusion with electrical permittivity (equation 2.3).

the same as that appearing in equation 4.1, i.e. the piezoelectric strain coefficient. Figures 4.1 (c) and (d) illustrate this converse effects for tensile strains. It is seen from equation 4.2 that the unit of d is the $\mathrm{m\,V^{-1}}$ but, from a consideration of Joule's law it can be shown it is the same as $\mathrm{C\,N^{-1}}$.

As a result of the anisotropy of piezoelectric materials, the quantities expressed in equations 4.1 and 4.2 are dependent on direction. Furthermore, the application of stress and field along, say, the x-direction product effects not only in this direction but also in the y- and z-directions as a result of the effects associated with the Poisson ratio. The stresses and strains each have components h from 1 to 6: 1, 2 and 3 for tensions and 4, 5 and 6 for shears, each in the x-, y- and z- directions, respectively, and the fields and polarizations have components i from 1 to 3 in these respective directions. Thus equations 4.1 and 4.2 may now be expressed in more detail as follows:

$$P_i = d_{ih}\sigma_h. \tag{4.1a}$$

$$S_h = d_{ih}E_i. \tag{4.2a}$$

Thus a d coefficient expressed as d_{11} evaluates the tensile strain in the x-direction produced by a field in the same direction: E_1. However, d_{12} evaluates the tensile strain in the y-direction and d_{15} the shear strain in the plane perpendicular to the y-direction as a result of this field.

The efficiency of a piezoelectric transducer, for a given direction, is expressed by the **electromechanical coupling coefficient**, k_c:

$$k_c = d(q/\varepsilon)^{1/2} \tag{4.3}$$

where ε is the electrical permittivity and q the appropriate elastic constant. k_c^2 can be shown to equal the ratio of the mechanical energy output to the electrical energy input and also the fractional decrease in the electrical permittivity of the transducer material because of the piezoelectric effect.

The piezoelectric characteristics d and k_c vary with temperature, generally decreasing with temperature rise to zero at what is called the **Curie temperature**. Exceptionally. for some ceramic transducers, the values of d and k_c increase from zero at a lower Curie temperature to a maximum and then decrease back to zero at an upper Curie temperature. The materials of these transducers can be injected with impurities to provide optimum performance at a required temperature. Table 4.1 indicates typical values of d, k_c and upper Curie temperature for some transducer materials.

4.2.2 Dynamic characteristics of piezoelectric transducers

Piezoelectric materials of suitable shapes such as discs and slabs with their faces coated with electrodes, can be set up into forced oscillations

Table 4.1 Approximate values of *d* coefficients, electromechanical coupling coefficients *k* and upper Curie temperatures T_c for piezoelectric transducers excited in the thickness mode

Material	d ($CN^{-1} \times 10^{12}$)	k	T_C °C
Barium titanate	60–190	0.2–0.5	120–140
Lead meta-niobate	85	0.42	550
Lead zirconate titanate (PZT)	80–320	0.23–0.76	250–490
Lithium niobate	6	0.18	1210
Quartz (X-cut)	2.3	0.11	350
Sodium meta niobate	86–160	0.51–0.53	290–420

at a given frequency by applying an alternating voltage. They can also be made to undergo free damped vibrations by shock excitation, typically by applying a direct current (DC) 'spike' of from several hundred to a thousand volts to the electrodes (section 2.2.2). In the receiving mode, alternating equal and opposite charges are induced on the electrodes as a result of the vibrations caused by the ultrasound. The characteristics of these vibrations can be determined either by measuring the current, I, through the transducer, which acts as an electrical impedance (as shown below), or the voltage, V, across the electrodes. At any given time, V is related to the mechanical force F, acting on the transducer, by:

$$F = pA = \alpha_T V, \tag{4.4a}$$

where p is the acoustic pressure and A the area of the surface exposed to the ultrasonic field. I is related to the velocity u of the transducer surface by:

$$I = \alpha_T u, \tag{4.4b}$$

α_T is called the **transformation constant**, which can be shown to be equal to Adq/l, where l is the thickness of the transducer and q the appropriate elastic constant, i.e. Young's modulus because the stresses are uniform over the whole cross-section. α_T should not be confused with the transmission coefficient α_t (section 2.5.1). One can compare the vibrations of a piezoelectric transducer with those of a mass connected to a fixed support by a spring (section 2.2), bearing in mind that the mass and spring now form part of a lumped system: they are no longer separate units. As before, the vibrations are analogous to the electrical oscillations of an inductance L (c.f. mass M) in series with a capacitance C (c.f. compliance C_m) and a resistance R which is equivalent to the mechanical resistance R_m of the transducer. Thus the mechanical impedance Z_m is given by:

$$Z_m = R_m + j(\omega M + 1/\omega C_m). \tag{4.5a}$$

Because $Z_m = F/u$ and the electrical impedance $Z = V/I$, we have, from equations 4.4a and b:

$$Z = V/I = Z_m/\alpha_T^2. \tag{4.5}$$

The equivalent electrical circuit (Fig. 4.2) consists of M, C_m and R_m in series with the secondary coil of a notional transformer having a turns ratio equal to α_T. The capacitance C_0 shunted across the primary coil of the 'transformer' is the electrical capacitance of the transducer. In the absence of the piezoelectric effect, this is a pure capacitor, assuming negligible dielectric losses in the transducer material. M and C_m have only 'effective' values because, as a result of stationary wave formation (section 2.6), spatial variations in vibration amplitude occur within the transducer. At the fundamental frequency of a transducer with unclamped surfaces (at half-wavelength resonance) the particle velocity antinodes occur at the surfaces and the nodes midway between them. It can be thus easily shown that, at this frequency, the effective mass is equal to one half of the actual mass.

The fundamental resonance frequency, f_0, at which the output of the vibrating mass is a maximum, is equal to $1/2\pi(MC_m)^{1/2}$ (section 2.2.4), where the thickness, l, is equal to one half wavelength, $\lambda/2$, and taking equation 2.18 into consideration, we have:

$$f_0 = c/\lambda = c/2l = (1/2l))(q/\rho)^{1/2}, \tag{4.6}$$

where q is the appropriate elastic modulus, c the velocity of sound in the direction of the required polar axis, λ the wavelength and ρ the density of the material of the transducer. The speed of compression waves in most commonly-used transducer materials is of the order of 5000 m s^{-1} and it can thus be seen from the above equation that the resonance frequency of a transducer 1 mm thick is about 5 MHz. Equation 4.6 shows also that the thickness varies inversely with frequency. Difficulties may arise in accurately cutting transducer materials

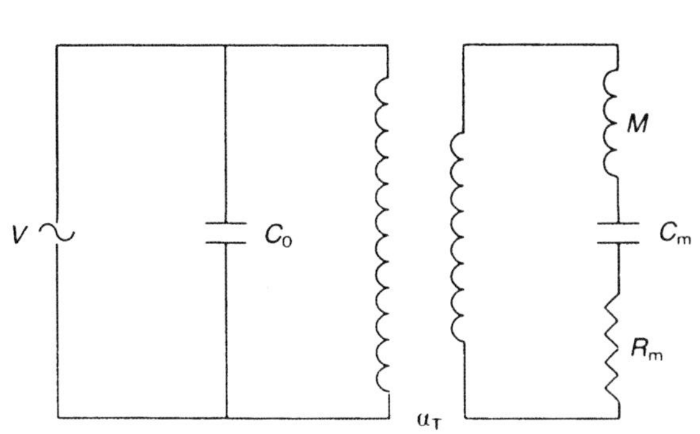

Fig. 4.2 Equivalent electrical circuit of a piezoelectric transducer.

to provide the very small thicknesses characteristic of fundamental frequencies of 10 MHz and above. With very thin transducers, there is a risk of discharge across the electrodes.

The generation of higher frequencies, above 10 MHz, can, however, be obtained with thicker transducers operated at upper harmonics, such as a 1 mm thick 5 MHz transducer at its 5th harmonic to produce 25 MHz. In this instance, because of the necessity for charges of opposite polarity to appear at the opposite surfaces, only odd harmonics can be excited (Blitz 1967). As an alternative, it is possible under certain circumstances to coat a thin layer of transducer material to a solid test object by means of a high-vacuum technique but this method is more appropriate to work in a research laboratory.

An important characteristic of a piezoelectric oscillator is its mechanical quality or '*Q*' factor (section 2.2.4) which varies inversely with the frequency bandwidth (equation 2.10). Its value decreases with the mechanical resistance R_m, which comprises the load provided by the specific acoustic impedance Z_a of the material of propagation, and also the damping characteristics of the transducer itself and any externally applied mechanical damping. In addition to possessing high electromagnetic coupling coefficients, ceramic transducers have heavy damping properties which give rise to the propagation of short pulses with wide frequency bandwidths. As a result, with the aid of suitable filters, it is possible to detect narrow bands of ultrasound over a wide frequency range.

4.3 THE PROPERTIES OF PIEZOELECTRIC TRANSDUCERS

4.3.1 General considerations

Table 4.1 enumerates the characteristics of a number of piezoelectric materials used as transducers. In general, the naturally occurring single crystal piezoelectric materials, such as quartz, have relatively low values of d and electromechanical coupling coefficient (k_c). They also have high electrical impedances and thus require relatively high voltages for adequate excitation. For example, at a frequency of 2.5 MHz, a 10 mm diameter quartz crystal transducer has an impedance of 20 kΩ but the impedance of a lead meta-niobate transducer of the same diameter is only 100 Ω. However, quartz transducers have comparatively high '*Q*' factors and can be used to advantage for measurements over narrow frequency bands, such as of frequency-dependent attenuation coefficients.

4.3.2 Ceramic transducers

Ceramic transducers are highly suitable for ultrasonic pulse-echo flaw detection in view of their properties discussed earlier: low impedance, low '*Q*' factors and high values of d and k_c. The materials used for them are not piezoelectric in their original states but possess a property called **electrostriction** which enables a mechanical strain S to be produced by the application of an electric field E, but there is no linearity between E, and S. Furthermore, S does not change in sign when the direction of E is reversed (c.f. magnetostriction, section 4.5.3) and varies in time in the manner of an unsmoothed rectified AC signal which would provide vibrations having a fundamental frequency equal to twice the natural frequency of the transducer and with a large number of harmonics. However, if a permanent DC polarizing field of sufficiently high intensity is suitably applied, a corresponding residual strain is produced. The application of an alternating field will thus increase or decrease this strain in a periodic manner and, provided that the amplitude of the latter is considerably less than the residual strain, the variations of field and strain are linear to a high degree of approximation and equations 4.1 to 4.5 are applicable: piezoelectricity is artificially induced in the transducer material.

A ceramic transducer can be initially polarized by heating it in an oven to well above the upper Curie temperature and then allowing it to cool whilst a direct voltage of the order of 1 kV is applied across the electrodes; this voltage is removed when the cooling is completed. Virtually all ceramic transducers are polarized in the thickness direction so as to enable compression waves to be transmitted and received. Exceptionally, for direct shear wave propagation, as obtained with a *Y*-cut quartz crystal, the polarizing voltage is applied perpendicularly to the thickness direction although the exciting voltage is still applied in the thickness direction.

The properties of a ceramic transducer can be varied by the injection of small amounts of impurities and, in this way, the d coefficients and Curie temperatures can be varied over wide ranges (Table 4.1). It should be mentioned that, although ceramic transducers are not, strictly speaking, crystals, they are often referred to as such.

4.3.3 Bimorph transducers

When a body is subjected to a tensile stress, it experiences not only a positive strain in the direction of the stress but also negative strains in directions at right angles to it. The values of the latter depend on the Poisson ratio. Thus, when a transducer in the form of a rectangular plate is energized in this way, resonances corresponding to the lateral dimensions are excited and the resulting frequencies are generally much lower than the fundamental frequency for the thickness mode. This phenomenon is exploited with the design of **bimorph** transducers which are used for low frequency operation: 25–40 kHz. A bimorph transducer consists of two identical, thin, square piezoelectric plates, with their polar axes for compression waves at right-angles to the thickness direction. They are glued together face-to-face with the polar axes perpendicular to one another. The electrodes are plated on two opposite ends of the combination instead of on the faces. Consequently, when a potential difference is applied across these electrodes, one plate expands and the other contracts in the direction of the electric field and the device, as a whole, executes high-amplitude flexural oscillations at a resonant frequency for a cross-sectional dimension. The transducer also acts as a sensitive receiver and it has been used with air coupling for testing solids at low frequencies where the attenuation coefficient in air is effectively negligible (section 10.5.3). Rochelle salt, which is very cheap but unsuitable for conventional ultrasonic testing (because of its susceptibility to moisture) is often used for these transducers.

4.3.4 Piezoelectric plastic film transducers

Mention should be made of the discovery by Kawai (1969) of piexo-electric plastic film transducers, **electrets**, which are transducers made from polymers in which piezoelectricity is induced artificially by permanent polarization, as with ceramic transducers, but only after they have been **frozen**. They have comparatively low *d* coefficients and are prepared in the form of diaphragms rather than resonating devices. They have the additional advantage of non-brittleness, which enables them to assume the shapes of spherical and cylindrical surfaces and thus to allow beam focusing without the need for lenses.

An account of electrets made from polyvinylidene fluoride (PVDF), which has a d coefficient of $20 \times 10^{-12}\,\mathrm{C\,N^{-1}}$, has been given by Murayama *et al.* (1976).

More recently, Rost *et al.* (1991) describe the use of a more effective electret material, polyvinylidene fluoride trifluoroethylene (P(VDF-TrFE)), available with thicknesses ranging from 6 to 100 µm and allowing the propagation of frequencies of up to 100 MHz.

4.4 PIEZOELECTRIC ULTRASONIC PROBES

4.4.1 General considerations

For the majority of ultrasonic testing applications, piezoelectric ceramic transducers are mounted in assemblies called **probes**. For the production of short pulses, as usually required for flaw detection and thickness gauging, the probes are shock-excited with a DC source of several hundred volts so as to perform heavily damped vibrations over a broad band containing the fundamental resonance frequency. For accurate time-delay measurements, with a conventional equipment, a sharp rise in the leading edge of the pulse is essential and it is often specified that the first peak should be reached before the first quarter-cycle is completed. These probes can also be used for narrow-band measurements if the received pulses are suitably filtered.

The term **probe** is used to describe the complete assembly consisting of the transducer itself, damping 'slug', mounting and containing case. Figure 4.3(a)(i) illustrates the construction of a typical compression-wave probe used for flaw detection with the pulse-echo technique. The damping 'slug' is made from a highly attenuating material such as an epoxy resin and its upper surface is inclined at an angle to the lower surface so as to ensure that reflections of waves from the rear surface of the transducer are suitably diverted to have died down sufficiently to avoid detection and, hence, the appearance of spurious indications. The front surface of the transducer is provided with a cover, originally fitted to protect it from wear and contamination during surface-scanning. With modern probe designs the cover would be made perhaps from a hard-wearing ceramic such as alumina with a characteristic impedance of a value lying between that of the transducer and that of the test sample; this enables an increase of the transmission coefficient (section 2.5.1), i.e. improved impedance matching.

It is usual practice for a single probe to act both as a transmitter and receiver but it may be necessary to use separate transmitter and receiver probes (section 5.3). Figure 4.3 (a)(ii) illustrates the design of a double or **twin crystal** probe in which a separate transmitter and receiver are located in the same housing. In this instance, it is important to ensure that there is no lateral ultrasonic transmission between them as a result of the effects of side lobes (section 2.7). This can be done by mounting plastic buffers, or delay blocks, to the radiating surfaces of the two probes and then inserting a layer of highly attenuating material between them.

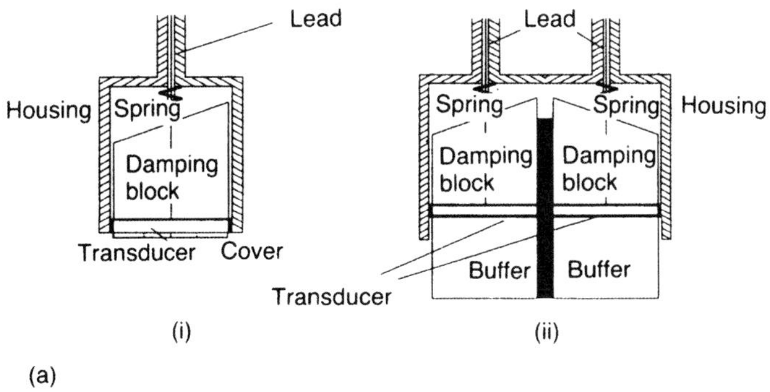

Fig. 4.3 (a)(i) Basic ultrasonic compression-wave probe; and (ii) twin compression-wave ultrasonic probe.

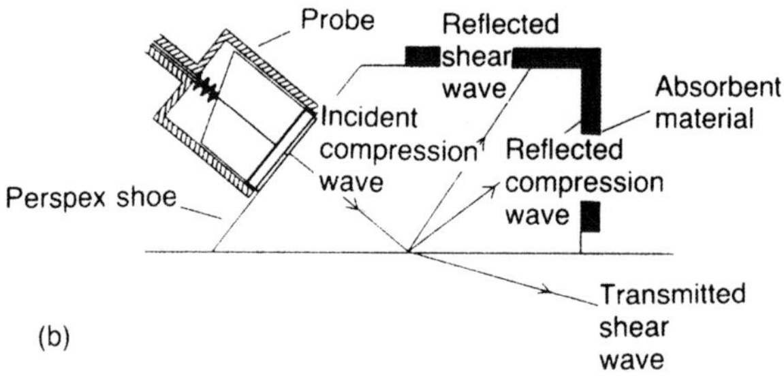

Fig. 4.3(b) Typical shear-wave probe arrangement.

Shear waves are usually transmitted and received by **angle probes** (Fig. 4.3(b)) consisting of compression-wave transducers mounted on Perspex wedges, often known as **shoes**, shaped in such a way that the angle of incidence at the interface between the wedge and sample under test is greater than the first critical angle (section 2.5.2). Because of mode conversion, only shear waves enter the sample. To avoid the possibility of the internally reflected waves in the wedge eventually finding their way into the sample after some time delay, as a result of multiple reflections, the vertical and top surfaces of the wedge are usually covered with plastic absorbent material.

Angle probes are manufactured with wedges having different angles, identified by the angle of refraction into a specified grade of mild steel. Because the speeds of compression waves in Perspex and shear waves in mild steel are approximately $2700\,\mathrm{m\,s^{-1}}$ and $3000\,\mathrm{m\,s^{-1}}$, respectively, the angle between the transmitted and incident waves is small. An angle probe marked with a value of 60° provides an angle of incidence equal to the angle of the wedge: approximately 51° (equation 2.48). If the same probe were to be used for testing an aluminium sample in which the shear wave speed is $3100\,\mathrm{m\,s^{-1}}$, the angle of refraction would then be 63°. Thus the probe angle should always be checked by means of a suitable test or calibration block before use (section 7.2).

Angle probes are also used for compression wave propagation provided that the angle of incidence is less than the first critical angle. They may be used in pairs to detect defects located just below the surface or singly for locating defects situated in what would be an 'off-beam' position with a normal probe, such as at the lateral surface of an axle tested in the axial direction. Care should be taken to identify any shear waves produced by mode conversion which would normally be of low amplitude for the small probe angles used (e.g. 11°). Compression wave angle probes may also be used for measuring crack sizes using the time of flight diffraction method (section 8.2.3) for which probe angles of 40–60° are used.

The design of a probe and the mounting of the transducer are highly critical operations. To achieve any degree of accuracy in measurements, it is essential that the axis of the ultrasonic beam coincides with that of the probe and that the beam is symmetrical about this axis. The characteristics of the beam can be detected by means of a Schlieren visualization technique (section 4.5.5) or more simply with a flaw detector by observing the amplitudes of signals received from a small metal sphere in the sound field. This requires that both the probe and reflector are suitably mounted in water to ensure proper acoustic coupling (section 4.4.2). With the latter method, contours indicating amplitudes in decibels, with respect to a standard level, are plotted as in Fig. 4.4, which indicates the existence of side lobes due to diffraction.

4.4.2 Coupling of piezoelectic probes

For maximum sensitivity to be achieved in ultrasonic testing, it is important to ascertain the highest possible degree of acoustic coupling between the transducer and the object being tested and this is ideally obtained when they are in intimate contact and their characteristic impedances are as near equal as possible. Equation 2.45 defines the **transmission coefficient** which indicates a decrease in acoustic intensity when the waves cross a boundary by an amount dependent on the values of these quantities. Equations 2.50(b) and (c) give values of transmission coefficients for transmitted compression and mode-converted shear waves, respectively, for compression waves in a liquid incident obliquely to the plane surface of a solid at different angles.

Generally, it is difficult to attain intimate contact between the surfaces

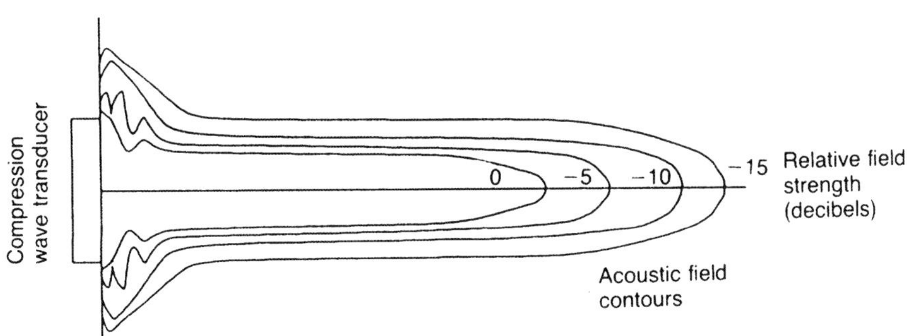

Fig. 4.4 Example of the configuration of the acoustic field in front of a compression-wave transducer. Numbers indicate relative levels of countours, in decibels.

of a transducer and a test object, unless they are optically flat, and there is effectively a very thin air gap between them, which could give rise to a very high transmission loss (equation 2.46) of the order of 40 dB or more. Taking account of two crossings of such a gap by the ultrasound having to pass both to and from the transducer during the test, the dynamic range of an ultrasonic set (about 100 dB) would rapidly be exhausted. However, with the use of a suitable liquid or semi-liquid couplant, e.g. a film of lubricating oil or grease, to fill the air gap the transmission loss may be reduced to only one or two decibels because the characteristic impedance of the couplant is of a magnitude comparable with that of the transducer and the test object.

When conducting a systematic scan, either manually or mechanically, over the surface of the examined object (section 5.6), introduction of air gaps may be difficult to avoid, and it may be necessary to use an immersion method for which both the probe and the object lie below the surface of a liquid, usually water, which effectively acts as a delay line or buffer (section 5.3.4 and Fig. 5.4(a)). It is essential that the lower surface of the probe lies below the water surface and that the values of the depths of the top and bottom surfaces of the objects are such that the time taken for the return transit of ultrasound between the probe and the bottom surface of the object is less than that taken for two return transits between the probe and the top surface of the object, otherwise confusing echoses would be observed. When total immersion is either undesirable or inconvenient, contact may be obtained by means of a suitably located water jets as shown, for example, in Fig. 10.14 which illustrates the use of a **wheel or roller probe**.

A disadvantage of having a couplant between the surface of the probe and the test object is that internal reflections take place within the couplant and, because its width is inevitably less than a pulse length, interference occurs and the shape of the pulse signal changes, thus requiring a small correction to be made to the position of the peak of the pulse echo (Mak 1991). The correction can be eliminated by using the first back-wall echo, instead of the transmission signal, as the point of reference because the same correction is applicable to all echoes. This problem would, of course, not arise with either immerison testing or the use of a buffer, provided the time delay is greater than that provided by the pulse.

Where the use of a liquid couplant is undesirable, for example when it may contaminate the material being tested, dry coupling may be possible with the use of a rubber or synthetic rubber delay line kept under sufficiently high pressure to squeeze out any trapped air and thus maintain good contact during scanning. In addition to having a high compliance, rubber has a characteristic impedance very close to that of water. However, it suffers the disadvantage of having a high attenuation coefficient in the megahertz range of frequencies. Much work has been done in the development of the properties of synthetic rubbers with a view to increasing characteristic impedance, in order to improve matching with both the transducer and test object, and to decrease the attenuation coefficient.

Billson and Hutchings (1993) have reported successful work on the use of a newly designed synthetic rubber as a delay line which could be made sufficiently long to avoid interference of internal echoes with the first front- and back-wall echoes from an aluminum sample 26 mm in thickness.

When it is not posible to employ either a liquid or solid couplant, the use of a non-piezoelectric transducer might be considered. However, it may well prove to be either inconvenient to install, difficult to use, inefficient in operation or very expensive, and it may be worth while to consider the use of an air-coupled ultrasonic probe. The attenuation coefficient in air at lower megahertz frequencies is about $1-2\,\mathrm{dB\,mm^{-1}}$ (i.e. $1000-2000\,\mathrm{dB\,m^{-1}}$), which may sometimes be acceptable, but reducing to $1-2\,\mathrm{dB\,m^{-1}}$ at lower kilohertz frequencies.

As indicated at the start of this section a loss of 80 dB may be involved with normal testing, but this can either be reduced or made acceptable with the use of properly designed probes, tuned at resonant frequency, high-input devices and highly-sensitive detectors. Improved impedance matching can be achieved, where possible, by mode conversion to shear or Rayleigh waves at the boundary with the test object because of a reduction of about 50% of characteristic impedance (ρc) compared with that for compression waves.

At one time, the use of air coupling was restricted to the low-frequency testing of low impedance materials such as plywood, with an upper frequency level of about 50 kHz. More recently (Rivera and Vitale 1988) a device has been designed for the through-transmission testing of sheets of composite material at a frequency of 400 kHz with air coupling and C-scanning (section 5.6). With this instrument, the transmitting transducer is precisely tuned to its resonant frequency for maximum output and the output of the receiving transducer is preamplified and turned a workable level. Improved resolution can be achieved by the use of focused probes. The device can be adapted to allow both probes to be located on the same side of the sheet to enable mode conversion into shear or Rayleigh waves.

A review summarizing techniques using dry coupling and non-contact coupling has been given by Szilard (1982).

4.4.3 Piezoelectric Composite Transducers

A recent development in transducer technology is the appearance of piezoelectric composite transducers in which artificially induced piezoelectric ceramics, such as PZT, are embedded into polymer resins (Gururaja *et al.* 1985). They can produce considerably higher outputs than those attainable by conventional transducers and are advantageous for testing highly attenuating materials and for achieving good air coupling. The ceramics can be made in a variety of shapes, such as rods, cubes and spheres and two arrangements are illustrated in Fig. 4.5. The opposite faces of these devices are, of course, covered with electrodes. The effect is to provide a reduced value of characteristic impedance and thus to improve matching with low impedance substances, such as air, and to increase electromechanical coupling.

Farlow and Hayward (1994) have experimented with transducers similar to the one illustrated in Fig. 4.5(a) but using ceramic pillars rather than cubes for ease in achieving the desired thickness resonances: a 20% ceramic/polymer volume fraction was obtained. With this arrangement it was possible for the electrical and mechanical resonances to take place at different frequencies, the values of which, for a given thickness, depend on the volume fraction. The transducer was found to be more efficient as a transmitter at the electrical resonance frequency and as a receiver at the mechanical resonance frequency. Optimum performance was achieved with the use of separate transmitting and receiving transducers and the possibility of an improvement of 100 dB in the signal-to-noise ratio has been claimed for this arrangement.

4.4.4 Piezoelectric transducer arrays

In the discussion of ultrasonic focusing (section 2.10), it was pointed out that ultrasonic lens systems and, hence, focused probes are sus-

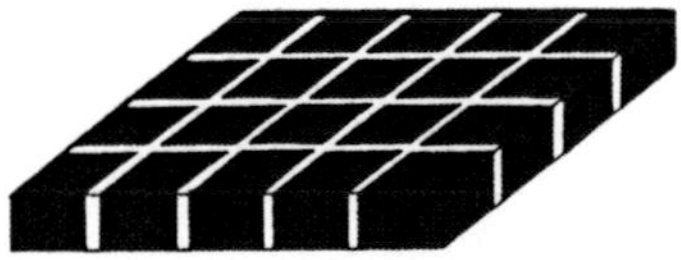

(a)

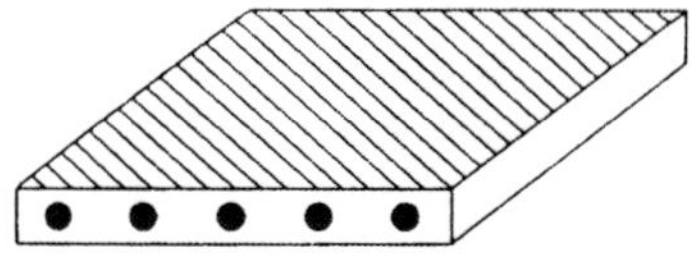

(b)

Fig. 4.5 Composite piezoelectric transducer arrangements (a) PZT cubes (dice) and (b) PZT spheres in polymer resin and composite. Reproduced by permission of Professor R.E. Newnham, The Pennsylvania State University.

ceptible to chromatic aberration and thus can produce blurred images. This might be tolerated in many instances but, if clear images are required, recourse should be made to the more sophisticated phased-array devices which eliminate aberrations and enable focal lengths to be varied at will. Phased arrays were originally used for radar systems and, with ultrasonic testing, they are normally used for scanning an immersed object so as to obtain a detailed image of structures and internal defects inside it. The arrays consist of a number of identical transducer sources placed in juxtaposition and excited periodically but at different phases to one another. By suitably adjusting the phase differences, beams from each of these sources can be made to focus to any desired point in the material under test and thus enable high resolution imaging of, say, internal defects in an object, Furthermore, unwanted parts of the image can be eliminated by filtering out all received signals except those from the single spot which is targeted by all components of an array.

Gebhardt (1983) describes three types of arrays:

- **annular**, producing rays concentric with the beam axis and having the advantage of symmetry but the disadvantage of a fixed beam of incidence;
- **linear**, consisting of a number of similar rectangular bars of similar thicknesses of the order of one wavelength in width and providing a two-dimensional field for a given position; and
- **planar**, having the elements distributed on a plane surface to allow a three-dimensional field.

Linear arrays are clearly highly suitable for ultrasonic scanning and their construction can be simplified by producing them in a monolithic form (Cochran *et al.* 1991): as a rectangular piezoelectric plate with the whole of its rear surface coated with an electrode and the front surface coated only in parallel strips of equal size and evenly spaced. Cochran and colleagues report the use of prototypes consisting of between 32 and 160 elements having widths ranging from 0.1 to 2 mm with centre-to-centre element separations of from 0.25 to 4 mm and operating at frequencies ranging from 1 to 20 MHz. With the exception of high frequency operation, the widths and separation distances could be changed as convenient.

Figure 4.6 illustrates the principles of the operation of a linear array of a parallel beam of ultrasound in a given direction: at an angle θ to the normal, to produce a plane wave front. The strips are located from one another at a constant distance, w, between their centres. If they are excited in turn at intervals of δt, the path difference between each neighbouring beam is given by the expression $c\delta t$ and it is clearly seen that the angle θ is given by:

$$\sin\theta = c\delta t/w. \tag{4.7}$$

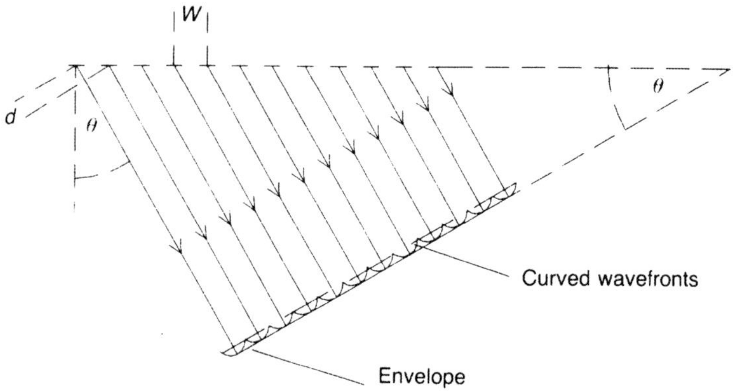

Fig. 4.6 Production of a parallel beam of ultrasound at an angle θ from a linear array consisting of equally spaced transducer strips having centres spaced apart by a distance w: d = path difference $c\delta t$ between neighbouring beams. The envelope of the curved wavefronts of individual beams gives rise to a plane wavefront. (Reproduced with permission from Silk 1984.)

Thus if $\theta = 30°$, $w = 4\,\text{mm}$ and $c = 1500\,\text{m s}^{-1}$ (the speed in water), the value of δt is 0.33 µs. To produce a beam converging to a focus at a given point, the value of the angle θ, and hence of δt, has to change from one strip to the next; i.e. the timings of the excitations must be suitably programmed to produce concave circular wave fronts.

4.5 OTHER TRANSDUCERS

4.5.1 Electromagnetic transducers

Electromagnetic transducers, sometimes called electrodynamic transducers when used at kilohertz frequencies, have long been used at audiofrequencies as loudspeakers and microphones and possess the advantage that the energy exchange between them and the medium of propagation takes place without the need for any material couplant. In their conventional forms, their uses are limited to lower ultrasonic frequencies because of the relatively high mass of their moving parts and the consequent drop in sensitivity with frequency. They have, nevertheless, proved highly effective as transmitters and receivers for resonance and damping capacity measurements (section 5.7.2) at frequencies of up to about 100 kHz.

The development of electromagnetic acoustic transducers (EMATs) has given rise to the use of electromagnetic propagation and reception of ultrasound at megahertz frequencies. This has arisen because the only 'moving parts' are the induced vibrations in the material being tested. The main restriction in the use of these transducers is that they can be used only for testing electrically conducting materials.

Electromagnetic transducers are reversible and depend for their operation on Fleming's rules of electromagnetism. When a current flows through an electrical conductor at right-angles to the direction of an applied magnetic field, Fleming's **left-hand** (or motor) rule postulates that a mechanical force is exerted on a conductor in a direction perpendicular to the plane containing the current and the field, thus causing the conductor to move. The directions of these quantities can be determined by observing directions of the second finger, forefinger and the thumb of the left hand, representing those of current, field and motion (force) respectively, when they are outstretched in mutually perpendicular directions. With a direct magnetic field and an alternating current, the conductor vibrates at the frequency of the current.

Conversely, Fleming's **right-hand** (or dynamo) rule states that an electric current is induced in a conductor in a direction perpendicular to the plane containing an applied magnetic field and force (giving rise to motion). Here the directions are indicated by the outstretched second finger, forefinger and thumb, respectively, of the right hand. Mechanical vibrations thus give rise to an alternating current, of the same frequency, in the coil with a direct magnetic field.

Electrodynamic transducers are extensively used for resonance and damping capacity methods of testing solid materials in bar form, especially samples of concrete (section 10.2) and an arrangement similar to that illustrated in Fig. 4.7(b) is used with electrostatic transducers. Longitudinal vibrations are excited although modifications can be made if flexural vibrations are to be studied.

Of interest is the work done by Barone and Giacomini (1954) on damping capacity measurements (section 5.7.2) from which the attenuation coefficient of a bar of dielectric or semiconducting material is determined at a frequency of up to 50 kHz. The bar is supported at its centre by metal knife edges and a thin conducting film, of negligible mass is coated on the opposite halves of the upper and lower surfaces and the connecting end-face (Fig. 4.8). A horizontal magnetic field is directed at right angles to the coated end-face. When a current flows through the strip, the free end of the bar is displaced in an axial direction. Conversely, the motion of the bar in that direction induces a current in the strip. In this way, when an alternating current passes through the strip under the influence of a magnetic field, the bar vibrates. On the other hand, if the bar is set up into vibrations as a result of some external agency, an alternating current is induced in the strip.

EMATs now in use are capable of generating and receiving either compression or shear ultrasonic waves in metals at megahertz

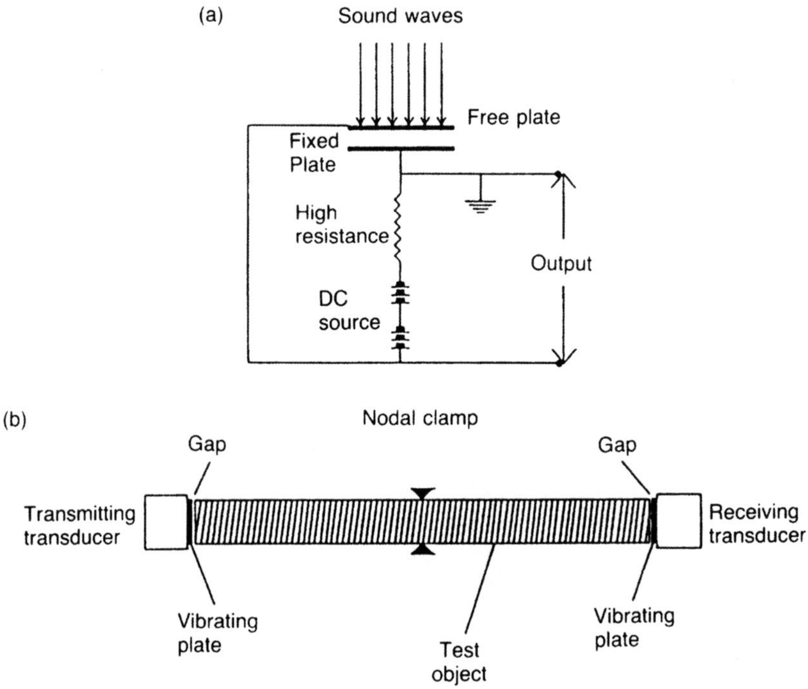

Fig. 4.7 (a) Basic design of an electrostatic transducer. (b) Arrangement for damping capacitance testing at resonant frequency using electrostatic transducers. (Based on information provided by CNS Electronics Ltd.)

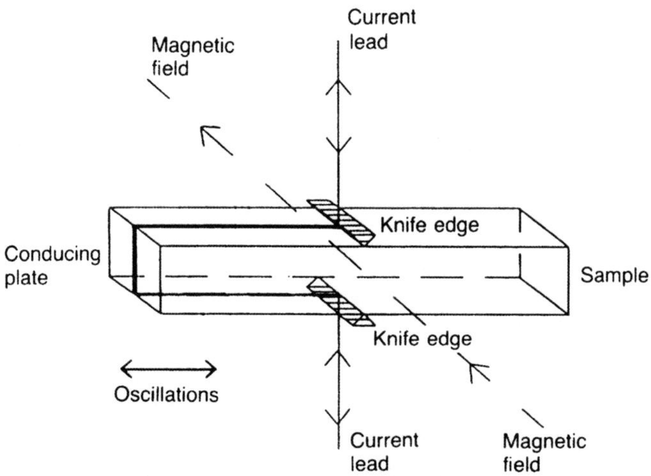

Fig. 4.8 Use of an electromagnetic transducer for damping capacity measurements at resonant frequency.

frequencies. Eddy currents are induced in the object being tested by an alternating current in a coil placed over and parallel to the surface which is under the influence of a direct magnetic field and ultrasound having the same frequency as the eddy currents is propagated. The ultrasonic polarization is determined by the direction of the magnetic field, in accordance with Fleming's left-hand rule explained above: compression waves with the field perpendicular to the intended direction of propagation and shear waves with the field parallel with it. The transducer, when acting as a receiver, picks up the eddy currents induced by the ultrasound in accordance with Fleming's right-hand rule.

Parkinson and Wilson (1977) have employed EMATs for inspecting metal bars and plates at high temperatures with the pulse-echo method and Fig. 4.9 illustrates how they are used for both compression and shear waves. For compression waves, a powerful magnet induces a field just below the surface of the sample and parallel with it and the eddy currents are generated and received by a pancake coil. For shear waves, the test sample is located between the pole pieces of the magnet. Because there are no moving parts in the transducer itself, the upper frequency limit is determined by the design of the eddy-current coil and the attenuation coefficient of ultrasound in the metal. Operation was achieved with success at temperatures of up to 900°C. Higher amplitudes are possible in ferromagnetic metals, such as mild steel, because of their ability to contain high magnetic flux densities. The amplitudes reduce with rise in temperature to minimum values at which the ferromagnetic properties disappear, i.e. at the Curie temperature, which is 750°C for mild steel.

Progress is being made towards reducing the physical sizes of EMATs by using small powerful permanent magnets such as those made from samarium cobalt, to produce the magnetic field and a flat spiral coil and such a device has been used by Hutchins *et al.* (1989) as a receiver in conjunction with a laser transducer source (section 4.5.4) propagating ultrasound into aluminium samples at high temperatures. Figure 4.9(c) illustrates an example of such a device consisting of a capsule having a 25 mm diameter and a height of 38 mm. Here the magnetic pole has a diameter of twice that of the coil. At present, it does not appear that permanent magnets are capable of producing the high field strengths attainable with electromagnets.

EMATs have now been considerably simplified in design and have the advantage over piezoelectric transducers of the lack of need for any coupling medium but, in general, they are lacking the versatility of the latter. Furthermore, they need to be located at close proximity to the object being tested. For remote testing, the more expensive laser transducers are suitable.

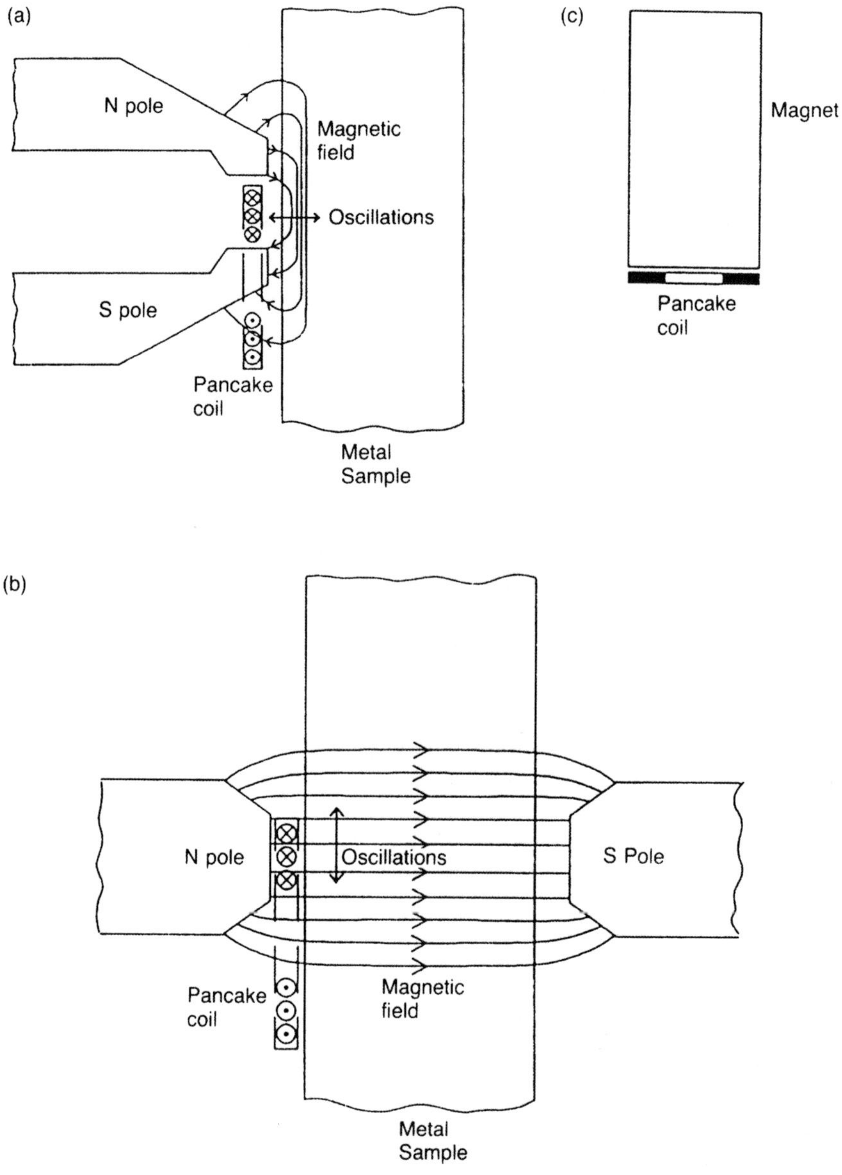

Fig. 4.9 Electromagnetic transducers (EMATs) for: (a) compression waves; and (b) shear waves. (After Parkinson and Wilson (1977)); (c) View of a typical Compact EMAT.

4.5.2 Electrostatic transducers

An electrostatic transducer is reversible and is essentially a capacitor comprising two parallel metal plates separated from one another by a very short distance (Fig. 4.7(a)). One of the plates is stationary and the other is free to oscillate in the thickness direction. When a direct potential difference is applied between the plates, equal and opposite charges appear on their surfaces and the resulting attractive force brings the free plate to a position of equilibrium nearer to the fixed plate. On applying an alternating potential difference, having an amplitude of less than the direct potential difference, the free plate vibrates periodically at the same frequency as that of the alternating voltage. When used as a receiver, a **capacitance microphone**, ultrasound is incident at the free plate, the latter vibrates and the capacitance and, hence surface charge, varies periodically at the wave frequency and thus gives rise to a flow of an alternating current having the same frequency as that of the ultrasound.

Because all points in each of the plates vibrate in phase with one another at the same amplitude, the electrostatic transducer is not a resonating device. It can thus be operated over a wide and continuous frequency range having an upper limit determined by the inertia of the vibrating plate and by the shunt capacitance to earth of the electrical leads. Using conventional parallel metal plates this limit is about 250 kHz but if the moving plate is either the end surface of a metal object being tested or a thin metal film coating the end surface of a non-metal object, this limit can be exceeded.

Of particular interest are measurements of resonance and damping capacity on materials in the form of bars mounted at their mid-points, so as to vibrate naturally at half-wavelength resonance in the longitudinal mode (Fig. 4.7(b)). The instrument is a version of ERUDITE (CNS Electronics Ltd, London) adapted for fitting electrostatic transducers. The transmitter is located at one end of the bar and the receiver at the other. The air gap between the vibrating plate of each transducer is only 0.03 mm and, for an operating frequency of 100 kHz, the value of kl in equation 2.46 is approximately 0.1, assuming a sound velocity of $340\,\mathrm{m\,s^{-1}}$ in air. This is sufficiently small for cos kl to approximate to 1 and sin kl to zero in equation 2.46 so that the transmission loss approximates to the acceptably small value obtained when the transducer and the bar are in close contact with one another (as represented by equation 2.45).

4.5.3 Magnetostrictive transducers

Magnetostriction or the **Joule effect** occurs in ferromagnetic materials such as nickel, cobalt and iron and also in ferrites. If a magnetic field is applied in an axial direction to a rod made from one of these materials, there is a mechanical tensile strain which is either positive or negative, depending on the nature of the material, and having a magnitude related to the physical state of the material and the strength of the applied field. When the direction of the field is reversed, there is no corresponding reversal of that of the strain: the sense of the strain is independent of that of the field. To obtain a sinusoidal variation of strain, a direct polarizing field having a magnitude greater than the applied alternating field is necessary, as with polarized ceramics when simulating piezoelectric transducers (section 4.3.2). In this instance, however, unlike a polarizing direct electric field, the polarizing magnetic field disappears when the polarizing device, such as a coil carrying DC or a permanent magnet, is removed.

The direct **magnetostriction coefficient** is defined as the fractional change in length for a given value of applied magnetic field and it has a high value for nickel compared with that for other materials. The converse magnetostrictive effect, the **Villari effect**, appears when the application of a tensile stress in the axial direction gives rise to a corresponding change in magnetic flux. In the presence of a polarizing magnetic field, the flux density varies linearly with the applied stress. Magnetostrictive transducers are reversible and a single device can act as both transmitter and receiver with equal efficiencies. The amplitude of the received stress, the acoustic pressure p, can be determined by measuring the induced current through the coil. As with piezoelectricity and electrostriction, the magnetostriction coefficient decreases with rise of temperature to the Curie temperature, at which the material is no longer ferromagnetic.

There are two kinds of magnetostrictive transducers:

- The 'traditional' indirect device comprises a ferromagnetic object, usually in the shape of a rod, which is set into vibrations which are transmitted to the material being tested. The upper frequency limit is determined by the dimensions of the rod, normally excited at resonance, and is about 100 kHz.
- A coil which can induce ultrasonic waves directly into a ferromagnetic or ferrimagnetic material which itself is being tested. Upper frequencies lie in the megahertz range.

Figure 4.10 illustrates the basic design of the first of these types of magnetostrictive ultrasonic transducer. A ferromagnetic rod clamped at a node is surrounded by two coils, one providing a polarizing field when a direct current, i_D, flows through it and the other exciting a sinusoidal magnetic field using an alternating current of amplitude i_0 (less than i_D) applied at the required frequency. The rod vibrates longitudinally and the maximum amplitude of the vibrations is obtained at the fundamental natural frequency, f_0, when its length, l, is equal to one half-wavelength:

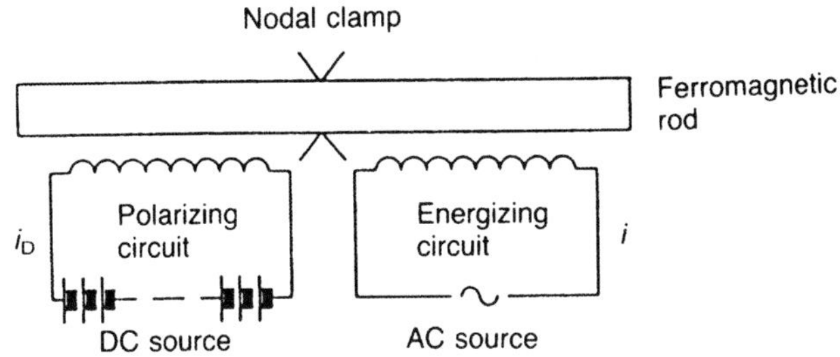

Fig. 4.10 Basic design of a magnetostrictive transducer line.

$$f_0 = (1/2l)(E/\rho)^{1/2}, \quad (4.8)$$

where E is the Young's modulus in the axial direction and ρ the density of the rod.

Given that the speed of compression waves in nickel is about 4800 m s^{-1}, the fundamental resonant frequency for a rod 125 mm in length is approximately 19 kHz. The output and sensitivity of this kind of magnetostrictive transducer are similar to those of ceramic piezoelectric transducers. However, because of difficulties in obtaining suitable coil windings for short specimens, the uppermost practical frequency limit is about 100 kHz, which is approximately the fundamental resonance frequency for a 25 mm long nickel rod. Generally, their more complex construction and higher manufacturing costs render them less advantageous than piezoelectric ceramics for ultrasonic work, even for high power and low frequency operations.

The second type of transducer is more suitable for testing rods and wires made from ferrous materials. It was originally devised for calibration of ultrasonic flaw detectors. The test material (Fig. 4.11) is polarized either by means of an encircling coil carrying a direct current or simply by stroking it with a strong magnet. Short pulses of AC are passed at regular intervals at the desired frequency through the transmitting coil which then induces compression wave pulses through

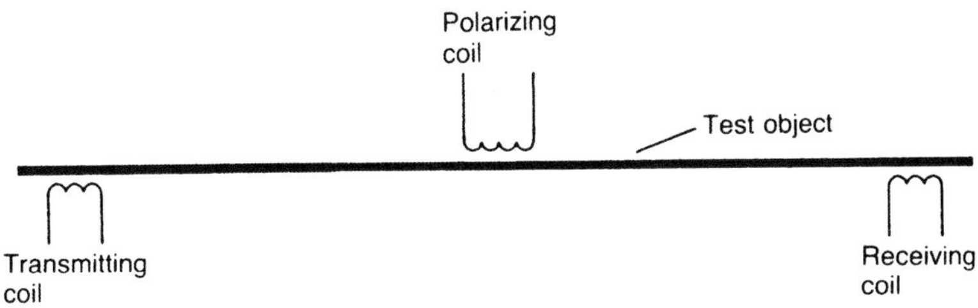

Fig. 4.11 Magnetostrictive transducer for megahertz testing of ferrous materials.

the test object. The pulses are picked up by the movable receiving coil where they are converted back into AC pulses. Alternatively, a single coil can be used as a reversible transducer to generate pulses and to receive echoes reflected at the end of the object. Because the length of the pulse is small compared with that of the rod or wire, no resonance takes place and the generation of megahertz frequency waves is possible. Very short pulses at megahertz frequencies can be transmitted and received with this transducer.

4.5.4 Laser transducers

The term 'laser' is an acronym for light amplification by simulated emission of radiation and describes a device for producing an intense monochromatic beam of coherent light or infrared waves by exciting either a solid rod, made from ruby or Nd-YAG (neodymium- yttrium aluminium garnet), or a gas or gas mixture. The frequency of the waves is characteristic of the nature of the excited substance. When used as a transmitting transducer, the laser beam, which is pulsed, is directed to the surface of the test object which can be located at a distance of up to several metres from it.

Laser ultrasonic transducers are particularly effective when testing metal objects because laser beams are electromagnetic by nature and hence are reflected at metal surfaces and cannot penetrate them by more than the standard penetration or 'skin' depth, δ, as defined by the depth at which the wave amplitude reduces to a fraction 1/e (e being the exponential function) of the surface value, equivalent to a drop of approximately 8.7 dB. It can be shown (Blitz 1991) that:

$$\delta = (2/\omega\sigma\mu_r\mu_0)^{1/2}, \tag{4.9}$$

where ω is the angular frequency of the laser beam, σ the electrical conductivity and μ_r the relative magnetic permeability of the metal and μ_0 the magnetic permeability of free space ($4\pi \times 10^{-7}\,\mathrm{m}^{-1}$). At the very high frequencies of the radiation from a laser, 10^{14}–10^{15} Hz, the skin depth is infinitesimal. For example, with a steel sample having an electrical conductivity σ of $10\,\mathrm{MS\,m}^{-1}$ and a relative magnetic permeability μ_r of 100, it can be seen from equation 4.9 that the standard penetration depth for a typical Nd-YAG infrared laser emitting at a frequency of 3×10^{14} Hz is of the order of 1 nm. Within this very small region, the energy of the electromagnetic radiation moving at a speed of $3 \times 10^8\,\mathrm{m\,s}^{-1}$ is absorbed in a time of less than about 10^{-15} s thus giving rise to shock thermal excitation.

If the laser emits a set of pulsed beams at the rate of N pulses at regular intervals in a time T, the resulting periodic shock thermal excitations are transformed into mechanical vibrations, as a result of expansion and subsequent contraction, and thus provide a source of ultrasonic radiation having a frequency of N/T: for 100 pulses generated at regular intervals during a period of 20 μs, the resultant frequency is 5 MHz but it is 250 MHz, for a period of 400 ns. With laser ultrasonic testing, these sets of pulses are repeated at regular intervals t and, for n sets, the pulse repetition rate of the resultant ultrasonic waves is equal to n/t: $50\,s^{-1}$ when $n = 50$ and $t = 1$ s. The shock excitations produced by the laser beam give rise to heavily damped vibrations which result in broad frequency bandwidths (section 2.2.4). These vibrations act as sources of compression, shear and Rayleigh surface waves in the test object.

Because the laser beams are very narrow, the diameters of the acoustic sources can be very small, a fraction of 1 mm, effectively providing point sources and thus a wide ultrasonic beam spread (section 2.7). The amplitude and direction of the emitted beam depends on the intensity of the laser beam. At low intensities, the vibrations of compression waves are orientated obliquely to the surface because, at the surface, there is insufficient resistance to constrain any forces acting at right angles to it. Figure 4.12(a) illustrates theoretical directivity curves (c.f. Fig. 2.12(b)) for (i) compression and (ii) shear waves for an aluminium sample. However, where feasible, coating the surface with a thin layer of grease, black paint or any other means of containment provides a sufficient amount of resistance to allow forward propagation of compression waves and a wider spread of directionality for shear waves (Fig. 4.12(b)).

With a sufficiently high intensity beam, enough energy is available

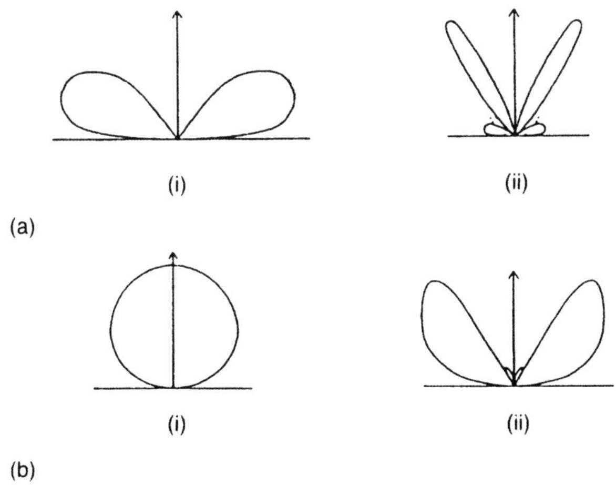

Fig. 4.12 Theoretical directivity patterns for a laser ultrasonic transducer propagating into aluminium for (i) compression waves and (ii) shear waves with (a) low intensity and (b) high intensity laser beams (after Scruby *et al.* 1982).

to provide the required amount of momentum to allow forward propagation of longitudinal waves to take place so that the directionality curves illustrated in Fig. 4.12(b) become applicable. However, this may give rise to ablation of the test object: damage to the surface as a result of the effects of very high temperatures, and test may be no longer non-destructive. This, of course, may not be a problem with, say, the testing of concrete, where the structure can tolerate some surface imperfection.

Laser generated ultrasound can be received by a transducer located on a suitable part of the surface of a test object but, if contact is possible, the use of a piezoelectric or other more conventional device should be considered. However, for remote testing, the use of an optical device is necessary for receiving. For receiving purposes, a low power continuous-wave laser simply scans the surface of the object in order to detect the ultrasonic vibrations originating from the source.

Taske *et al.* (1990) used, as the source, a Q-switched Nd-YAG infrared laser having a wavelength of 1060 nm transmitting 20 ns pulses producing compression, surface and shear ultrasonic waves of broad bandwidths with flat responses over the range 0–20 MHz. The receiver was a Michelson interferometer (Jenkins and White 1976) actuated by a He–Ne laser of 633 nm wavelength.

With the Michelson interferometer (Fig. 4.13), which is suitable for receiving low-amplitude compression waves, the laser beam is directed towards the surface of the test object and intercepted by a beam splitter (i.e. a semireflecting mirror). The undiverted part of the beam, after reflection at the surface, is then diverted at the beam splitter to combine with the reflection of that part of the beam, which was originally diverted, and reflected at the mirror. Interference occurs in the combined beam which is received by the detector after reflection at the

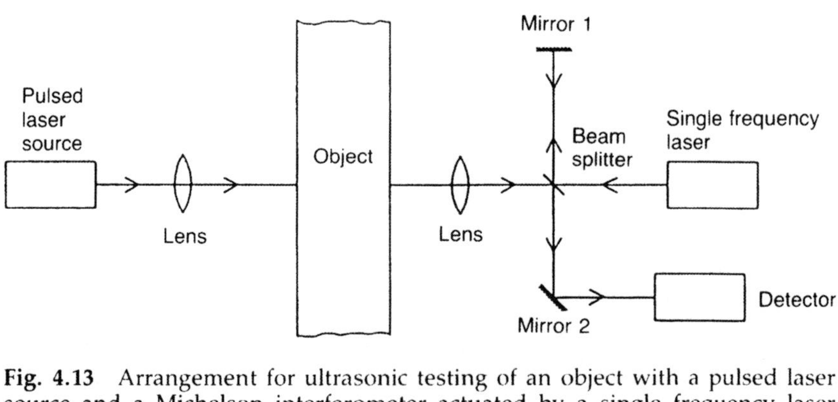

Fig. 4.13 Arrangement for ultrasonic testing of an object with a pulsed laser source and a Michelson interferometer actuated by a single frequency laser used for detection.

second mirror. The phase difference between the two interfering waves depends on the position of the surface of the test object, i.e. the phase of the ultrasonic vibrations at the surface. The variation of the amplitude of the interfering laser waves is similar to that of stationary ultrasonic waves (Fig. 2.10).

Equations 2.52 and 2.54 indicate variations of pressure amplitude and particle velocity for standing acoustic waves. The form of the equations for standing light waves is similar and, assuming a virtual 100% reflection at the solid boundary, i.e. $|r| = 1$, we have:

$$A = A_0 \sin kx = A_0 \sin(2\pi x/\lambda), \tag{4.10}$$

where A is the wave amplitude and A_0 the maximum amplitude. Equation 4.10 shows that the change of phase from maximum to zero amplitude (from maximum to zero optical intensity) is seen to occur for a shift of the reflecting surface by an amount $x = \lambda/4$ (one quarter-wavelength), as shown by the degree of intensity indicated by the photoelectric cell in the detector. It is impracticable to detect greater changes than this amount with any degree of certainty by this method. For the wavelength of 633 nm emitted by the He–Ne laser beam the value of $\lambda/4$ is 158 nm which represents the maximum ultrasonic particle displacement (section 2.3.2) that can be measured.

The typical values quoted above indicate that the interferometer method of detection is suitable only for very smooth surfaces. For objects having rougher surfaces, the Krautkrämers (1990) discuss a method in which the reflected laser beam is received directly by a photoelectric cell located behind a diaphragm. The light scattered by any surface irregularities gives rise to a speckle pattern which fluctuates with the motion of the surface; the corresponding output signals from the cell indicate the ultrasonic particle displacement amplitude.

The main advantages of testing by laser generated ultrasound are that it is a non-contact method, the transducers can be located remote from the object being tested, rapid scanning is possible, a wide range of frequencies can be used, there is no requirement for precise tranducer orientation and awkwardly-shaped objects can be examined. Disadvantages include the facts that there is a relatively low upper limit to wave amplitude, unless ablation can be tolerated, the equiment is expensive, bulky and requires highly-skilled handling and safety precautions are necessary in view of any potential danger of burns and eye damage.

4.5.5 Optical detection and field visualization

The use of coherent light in the form of lasers for transmitting and receiving ultrasound is described in the previous sub-section. Here, we are concerned mainly with the use of incoherent monochromatic light produced by sources such as filtered discharge tubes which are not practicable for use as ultrasonic transmitters but can be used to detect and visualize optical changes in a transparent object as a result of the passage of ultrasound through it and to relate them to the acoustic characteristics at different points. It should be pointed out that low-energy lasers can also be used here and are often found to be more suitable.

When sound waves pass through a transparent material, the resulting alternating strains are accompanied by periodic changes in the optical refractive index of the material. As a result, beams of light may be scattered, reflected, refracted, diffracted and undergoing amplitude and intensity changes. These factors are related to the speed of sound and the attenuation coefficient and their applications can serve as ultrasonic detectors.

Visualization of ultrasound makes it possible, for example, to study the characteristics of a beam of sound radiated by a piezoelectric transducer immersed in water and also the way in which the ultrasound is reflected and scattered by artificial defects in a transparent block of a material such as Perspex. An effective way of visualizing ultrasound is to use the Schlieren technique devized by Toepler in 1886. With the basic form of this method, a vertical slit (Fig. 4.14) is illuminated by a source of monochromatic light such as a mercury arc lamp covered by a suitable filter. The beam of light cuts through the sound field and is brought to a focus at a vertical wire or narrow strip of sufficient width to obstruct the beam when the sound is absent. The field of view of an optical instrument at the focus is then dark.

When the sound waves, either pulsed or continuous, are transmitted through the medium, the spatial variation of its refractive index causes deviations of the light beam so as to bypass the obstacle and thus enter

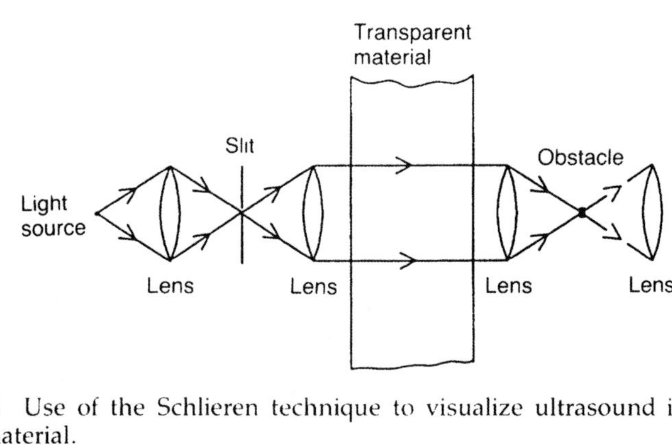

Fig. 4.14 Use of the Schlieren technique to visualize ultrasound in a transparent material.

the field of view of the observing optical instrument. These deviations take place in a periodic manner at the frequency of the ultrasonic waves and, if the source of light is interrupted at that frequency, a stroboscopic effect is observed and a pattern of light and dark fringes is seen. With plane waves, the fringes are parallel, but with spherical waves, they take the form of circular arcs. With continuous waves, the fringes cover the whole field of view but, with pulsed waves, the fringes cover the whole field of view but, with pulsed waves, they are limited by the widths of the pulses. Thus visual images of the waves are formed and, when the waves are irregular as a result of non-uniform scattering and diffraction caused for example by simulated defects such as saw cuts, images of the irregularities are seen. With stationary ultrasonic waves, parallel fringes indicating the nodes and antinodes (section 2.6.3) can be seen without the need for stroboscopic illumination.

As an alternative to the Schlieren technique, visualization of sound waves can be brought about by the photoelastic effect with which the mechanical strains characteristic of the ultrasonic waves give rise to intensity variations of polarized light illuminating the medium of propagation. Work has been done on the visualization of ultrasound by Hall (1977) and Archer Hall and Hutchings (1979).

The speed and attenuation of ultrasound in a transparent material can be measured by using **optical diffraction**. The periodic variation of refractive index caused by the passage of the waves form an optical diffraction grating (Jenkins and White 1976) which moves through the material at the speed of ultrasound. However, the latter is typically of the order of 10^{-5} less than the speed of light and the grating is effectively stationary with respect to the light passing through it, which does so at right-angles to the direction of the ultrasonic beam (Fig. 4.15). The angle θ of the diffraction of light incident at right-angles to the grating is given by the following expression:

$$\sin\theta = n\lambda/d, \tag{4.11}$$

where λ is the optical wavelength and n a whole number, either positive or negative, ranging from zero to an upper limit set by the condition that $\sin\theta$ must always lie within the range -1 to $+1$. d is the grating spacing which is equal here to the ultrasonic wavelength. It can be seen that a total of $2n + 1$ beams emerge from the grating.

The sample under test should, of course, be transparent and the sides perpendicular to the incident light beam should be accurately perpendicular to one another. For maximum resolution of the grating, it is important that the ultrasonic frequency band should be as narrow as possible and, hence, the use of short-pulsed waves is generally unsuitable. The source of light is usually a slit illuminated by a filtered discharge tube, although better results are attainable with a laser

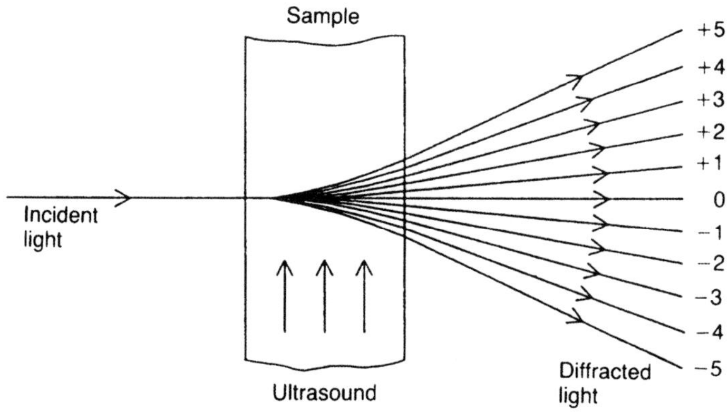

Fig. 4.15 Basic principles of the optical diffraction method; numbers +1, −1, +2, −2, etc. indicate orders of different diffracted beams of light, arrows indicate directions of ultrasonic and optical beams.

source. The beam from a slit is rendered parallel by means of a lens and it is this parallel beam which passes through the ultrasonic field. Figure 4.15 illustrates the paths of the diffracted beams from which images of the slit can be produced by means of a suitable focusing instrument, such as a telescope. With a laser, it is important that risks of direct viewing are absent and the use of a photoelectric detector is thus preferred. From a knowledge of the optical wavelength, which is generally supplied by the manufacturers of the source, and the angle of diffraction, the ultrasonic speed can be readily calculated from equation 4.11.

For maximum resolution of the diffracted beams, as high a frequency as possible should be employed, with a minimum of 5 MHz, and the use of a quartz crystal source would be of advantage because it can be used at frequencies of up to about 20 MHz in its fundamental mode and at much higher frequencies if harmonics are excited (section 4.2.2).

The measurement of wavelength and, hence, wave velocity with the optical diffraction method is a 'one off' event and, in practice, is independent of the length of the sound field. The use of the optical diffraction technique is thus possible even when stationary waves (section 2.6) are formed because the latter are observed only as a result of the integration of time; equation 4.11 then applies to both progressive and stationary waves. This is not the case with the measurement of attenuation, which is measured by observing values of the optical intensity of the 0th order beam by means of a photoelectric cell at two different positions in the sound field after moving the light beam laterally and applying equation 2.34(a). Now the optical intensity varies inversely with the ultrasonic intensity and, with the formation of stationary waves, the resultant ultrasonic intensity at a given point is the sum of the intensities at that point for the incident and reflected beams in both directions; this makes it very difficult to identify the intensity contribution from the original incident ultrasonic beam. Hence, it is important that there is no reflection at the rear surface of the object and this can be achieved by coupling a highly attenuating material to that surface (section 5.7.1). In practice, the attenuation coefficient is determined after calibration with a test sample having a known value of this quantity. A fuller description of this method has been given by Bergmann (1938).

Further information regarding optical generators and detectors, including laser transducers, has been provided by McIntire (1991), the Krautkrämers (1990) and Szilard (1982) and laser transducers only by Birnbaum *et al.* (1984) and Scruby *et al.* (1982).

4.5.5 Miscellaneous transducers

There do not appear to be any significant applications of other possible forms of ultrasonic transducers in non-destructive testing except, perhaps, the use of an impact device to propagate a sound containing a continuous range of frequencies ('white noise') some of which would extend into the ultrasonic spectrum. In a private investigation (unpublished) involving damping capacity measurements, the author (J.B.) made use of this method in which a steel ball was connected by a length of string to a fixed support. The ball was held in position by means of an electromagnet and released to describe a quarter circle and then to strike the vertical surface of the sample under test. The latter was a nodally mounted rod, having a fundamental resonance frequency of about 20 kHz, which was set into damped free vibrations on impact. A capacitor microphone acted as a receiver. In this way the attenuation per cycle was determined and related to structural properties of the samples.

REFERENCES

Archer-Hall, J.A. and Hutchins, D.A. (1979) *Ultrasonics*, **17**, 209–14.

Barone, A. and Giacomini, A. (1954) *Acoustica*, **14**, 182–4.

Bergmann, L. (1938) *Ultrasonics*, (English translation H.S. Hatfield). Wiley, New York/Bell, London.

Billson, D.R. and Hutchins, D.A. (1993) *Br J NDT*, **35**, 705–9.

Birnbaum, G. and White, G.S. (1984) Chapter 8 in, *Research Techniques in Nondestructive Testing*, Vol 7 (ed R.S. Sharpe), Academic Press, London, pp. 259–365.

Blitz, J. (1967) *Fundamentals of Ultrasonics*, 2nd edn. Butterworths, London, Section 3.2.1, pp. 43–4.

Blitz, J. (1991) *Electrical and Magnetic Methods of Nondestructive Testing*. Adam

Hilger, Bristol, Section 2.8.2, pp. 42–3.
Cochran, A., McNab, A. and Hayward, G. (1991) *Br J NDT*, **33**, 495–505.
Farlow, R. and Hayward, G. (1994) *Insight*, **36**, 926–35.
Gebhardt, W. (1983) Section 2.6 in, *New Procedures in Nondestructive Testing*, (ed P. Höller). Springer-Verlag, Berlin, pp. 151–62.
Gururaja, T.R., Schulze, W.A., Cross L.E. *et al.* (1985) *IEEE Trans Sonics Ultrasonics*, **32**, 481–98.
Hall, K.G. (1977) *Ultrasonics*, **15**, 57–69 and 245–52.
Hutchins, D.A., Hauser, F. and Bresse, L.F. (1989) *NDT and E*, **5**, 17–25.
Jenkins, F.A. and White, H.E. (1976) *Fundamentals of Optics*, 4th edn. McGraw-Hill, New York, pp. 271–3.
Kawai, H. (1969) *Jpn J Appl Phys*, **8**, 975–6.
Krautkrämer, J. and Krautkrämer, H. (1990) *Ultrasonic Testing of Materials*, 4th edn. Springer Verlag, Berlin, Section 8.6.
McIntire, P. (ed) (1991) *Nondestructive Testing Handbook*, 2nd edn, Vol 7, *Ultrasonics*. American Society for Nondestructive Testing, Columbus, OH.
Mak, D.K. (1991) *Br J NDT*, **33**, 344–6.
Murayama, N., Nakamura, K., Obara, H. and Segawa, M. (1976) *Ultrasonics*, **14**, 15–28.
Parkinson, G.S. and Wilson, D.M. (1977) *Br J NDT*, **19**, 178–84.
Rivera, O. and Vitale, F.A. (1988) *Materials Evaluation*, **46**, 614–15.
Rost, M., Splitt, G. and Warkowski, W. (1991) *Br J NDT*, **33**, 597–8.
Scruby, C.B., Dewhurst, R.J., Hutchins, D.A. and Palmer, S.B. (1982) Chapter 8 in, *Research Techniques in Nondestructive Testing*, Vol 5 (ed R.S. Sharpe). Academic Press, London, pp. 281–327.
Silk, M.G. (1984) *Ultrasonic Transducers for Non-destructive Testing*. Adam Hilger, Bristol, p. 90.
Szilard, J. (1982) Chapter 9 in, *Ultrasonic Testing* (ed. J. Szilard). Wiley London, pp. 381–409.
Taske, C.G., Milne, J.M. and Smith, R.L. (1990) *Br J NDT*, **32**, 355–9.

CHAPTER 5

The principles of ultrasonic testing

5.1 INTRODUCTION

Ultrasonic testing involves, either directly or indirectly, one or more of the following measurements: time of transit (or delay) of the waves, path length, frequency, phase angle, amplitude, impedance and angle of wave deflection (reflection and refraction). The methods of measurement make use of either pulsed or continuous waves.

Pulse methods have the advantage of being able to provide readily defined and accurately located reference points for transit times which can be observed on the screen of a cathode-ray oscilloscope. They include the pulse-echo, the pulse-transmission (or shadow) and the pulse-resonance techniques and form the majority of everyday ultrasonic testing operations. Continuous-wave measurements with progressive waves present difficulties except for highly attenuating materials and we are principally concerned here with stationary-wave methods which are used mainly to investigate resonance phenomena.

In addition to the treatment of the more conventional ultrasonic methods, this Chapter deals with the use of the ultrasonic goniometer (section 5.8), impedance measurements (section 5.9), surface and Lamb wave applications (section 5.10) and acoustic emission (section 5.11).

5.2 PULSE METHODS

In principle, the pulse technique is the simplest of all the ultrasonic testing methods and it is the one most commonly used. It consists basically of measuring the time taken for a short train of sound waves to move through a given distance. This distance can be determined if the speed of sound in the material is known or, if the distance be known, the speed of sound can be readily calculated. By measuring the relative amplitudes of pulses which have travelled different distances, one can determine attenuation coefficients. Observing the patterns of the received pulses can provide valuable information about the nature of a defect at which the pulses have been reflected and, perhaps, of the structure of the material being tested.

The **pulse-echo** method is the most commonly used ultrasonic pulse technique. Others include the **pulse-transmission** or shadow method and the **pulse-resonance** method. The latter has been found useful for setting up stationary waves in regions remote from the source, such as where there is an abrupt change in cross-section of a rod over a short distance.

5.3 THE PULSE-ECHO METHOD

5.3.1 General considerations

With the pulse-echo method, the time taken for sound pulses, generated at regular intervals, to pass through the object being testing and to return is measured. The returned pulses may be received either after a single reflection at a discontinuity of characteristic impedance (a boundary or a defect) or multiple reflections between the discontinuity and the front surface of the object. Applications include flaw detection, thickness gauging and measurements of velocity and attenuation. The advantage of this method is that access from only one side of the test object is necessary, provided that the rear surface is parallel to the surface where the transducer is placed. The pulse signal received from this opposite surface is called the **back-wall echo** and the observation of this echo is essential for thickness gauging and attenuation measurements; it also provides a useful reference for flaw detection. However, there are some applications such as the shear-wave testing for defects orientated at right angles to the front surface for which no back-wall echo is possible.

Objects suitable for ultrasonic testing can vary in dimensions from a few millimetres to several metres and, for a typical wave velocity of $5000\,\mathrm{m\,s^{-1}}$, the time elapsing for the return of a single back-wall echo varies correspondingly from about 1 to $1000\,\mu\mathrm{s}$. Electronic methods are necessary to measure such times. The basic design of the equipment used for the pulse-echo method is illustrated in Fig. 5.1. It is commonly known as an **ultrasonic flaw detector**, although it can be employed for other applications such as those mentioned above.

With this equipment, a trigger excites simultaneously a **time-base** control and a **pulse generator** at regular intervals, usually between 50 and 1000 Hz: the **pulse repetition frequency** (PRF). On excitation, the time-base control which is connected to the X-plates of a cathode-ray oscilloscope, causes the electron beam in the tube to be swept in a horizontal direction from left to right. The time of the sweep can be

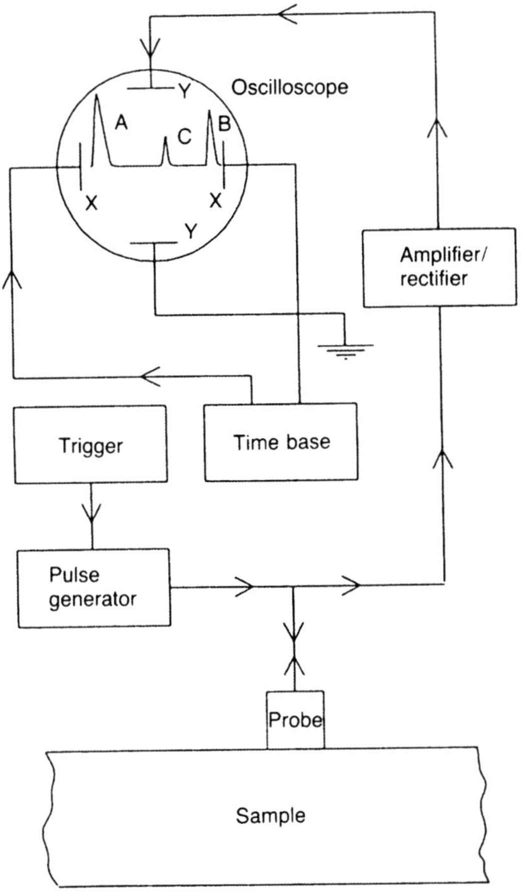

Fig. 5.1 Basic design of pulse-echo equipment (ultrasonic flaw detector); X = X-plate of oscilloscope, Y = Y-plate, A indicates transmission signal, B = back-wall echo, C = defect echo.

varied to suit the acoustic path length by means of a device known as the **range control**. This variation is continuous, and the range can be extended by several orders of magnitude using a multi-position switch. At the end of its sweep, the electron beam returns almost instantaneously to its original position in readiness for the next excitation of the trigger. In the absence of any signal applied to the Y-plates, a bright horizontal trace appears on the screen of the oscilloscope. It is often necessary to examine, in some detail, echoes at remote distances from the probe and this can be done by manipulating the **delay control** to shift, by a required amount, the zero of the time base and then expanding the latter, if necessary, by means of the range control.

On excitation by the **trigger**, the pulse generator feeds a direct potential difference, typically of from several hundred to a thousand volts and of almost instantaneous duration (a 'spike voltage') to the electrodes of either a transmitting or reversible heavily-damped probe (section 4.4) in contact with the object under test. The magnitude of this applied voltage can be adjusted by means of a **pulse energy control**.

For **single-probe operation** a reversible probe is used and this is connected through an amplifier and rectifier, which can be bypassed if a display of the waveform is required, to the Y-plates of the oscilloscope. The output of the amplifier can be adjusted by means of a sensitivity control. Thus, when the probe receives its shock excitation, it generates a short ultrasonic pulse and, at the same time, a large peak known as the **transmission peak** appears at the left-hand extremity of the oscilloscope display. Echoes of the pulses, which are received by the probe after reflection, are converted into electrical pulses and corresponding peaks appear on the oscilloscope screen.

For **double-probe operation** the transmitting probe is not directly connected to the oscilloscope in this way and, on transmission of the pulse, the transmission peak does not appear on the screen, although the output from the pulse generator can be designed to produce some form of time indication at its position, A. However the received echoes are picked up by the receiving probe, converted into electrical pulses and fed to the Y-plates of the oscilloscope in the same way as for single-probe operation.

Because of the finite amount of time taken for the sound pulses to travel through the material, the **time delay**, the displaced echo peaks are displaced to the right of point A by distances corresponding to the amounts of this delay. For example, the peaks B and C might correspond, respectively, to the first back-wall echo and an echo from an internal defect. Because of the rapidity of the pulse generation as determined by the PRF, the displayed peaks appear to be steady. The values of the time delays can be determined by calibrating the time base (section 5.3.2) but it is evident, assuming that the time base indications vary linearly with the horizontal distance, that the distance of the indicated defect from the front surface of the test object is equal to the product of its thickness and the ratio AC/AB. To facilitate the reading of distances along the time base and the heights of the echo-peaks, a graticule is fitted onto the oscilloscope screen. It contains a number of parallel lines of which one, usually on the lower part of the graticule, is graduated in millimetres from zero to 100. There is usually also a vertical scale, graduated in decibels, to assist with attenuation measurements.

It is often necessary to adjust either or both of the time base and

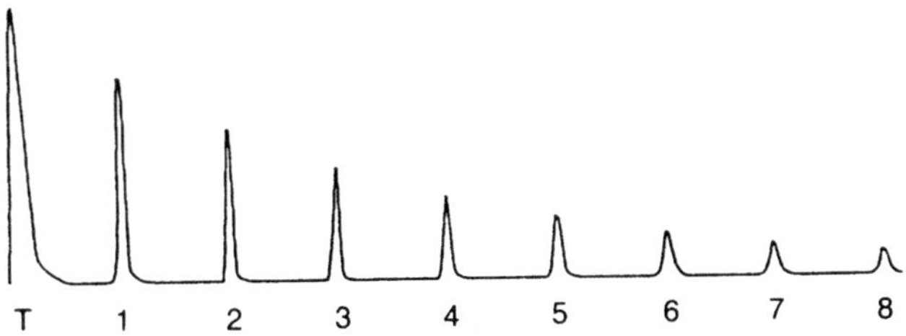

Fig. 5.2 Indications of multiple echoes on the screen of an ultrasonic flaw detector for a parallel-sided defect-free test-sample: T indicates transmission peak and the numerals indicate orders of back-wall echoes.

range control until a number of multiple back-wall echoes are displayed on the screen (Fig. 5.2). This may be necessary with velocity measurements and essential with attenuation measurements and time-base calibration with the aid of a test block.

For single-probe operation, account must be taken of a region, immediately in front of the probe, called the **dead zone** for which any echo indications are masked by the width of the transmission peak. It arises from the overloading of the amplifier by the shock excitation and a relatively long time may elapse before the signal reduces to zero. Thus defects situated in the sub-surface region may be undetectable with a single probe. The length of the dead zone can, of course, be reduced by decreasing the shock-excitation voltage by manipulation of the gain control but at the cost of reducing signal sensitivity. Alternatively, a delay medium can be introduced between the probe and the surface of the sample either by immersion or by inserting a Perspex block in front of the transducer to act as a buffer. This has the disadvantage of possibly introducing spurious signals arising from multiple reflections at the boundaries of the delay medium. The most satisfactory solution is to use a double probe.

5.3.2 Velocity and thickness measurements

The simplest method of measuring the speed of sound of compression waves in a parallel sided sample of known thickness, using the pulse-echo method, is to measure the displacement of the first back-wall echo from the start of the transmission peak. The speed of sound can then be calculated from the calibration of the time base. It should be remembered that each reflected echo represents a return transit through the material so that the acoustic path length is equal to *twice* the thickness of the sample.

For greater accuracy measurements are taken from the position of intersection of the leading edge of the pulse with the time-base line and not that of the peak* for the reasons given in section 2.9, especially if the material is highly attenuating. To facilitate this, it is preferable to use a probe giving a rapid rise time so that the edge is near as possible perpendicular to the time-base line. In practice, this is achieved by ensuring that the maximum amplitude is reached before the first quarter-cycle is completed.

The time base should be properly calibrated before any measurements are made and this includes checking its linearity. It is also essential to take into account the thicknesses of the probe cover, couplant and any buffer used. The IIW (International Institute of Welding) calibration block (section 7.2.2) is often specified for this purpose. When, for some reason, this block is not available, the calibration should be performed on a sample having known dimensions and speed of sound. Linearity can be checked by observing multiple echoes and measuring the separations of their leading edges. Allowance can be made for the effect of couplants etc. by deducting the time separation between the leading edges of two adjacent multiple echoes and subtracting this from the separation between that of the first echo and of the transmission peak.

If only the times of transit of the pulses need to be measured, the design of the equipment can be simplified by using a digital timer instead of the cathode-ray oscilloscope. The timing device is triggered at the time of the shock excitation of the probe and stopped by the return of the leading edge of the first echo, which can usually be detected more accurately than by visual means. An allowance should be made for the transit time through the probe cover and coupling fluid. Thus the speed of sound can be measured for a sample of known dimensions. Alternatively, after suitable calibration the instrument can be used to indicate dimensions as with the **digital ultrasonic thickness gauge** (section 6.3).

The measurement of the speed of shear waves using the normal pulse-echo method is somewhat more involved and the use of the goniometer (section 5.8) may be more suitable. However, for a solid in the form of a cylinder having a length sufficient to allow the diverging beam in the far field to be reflected from its sides (section 2.7), it is possible with the use of a compression-wave probe to measure the speed of shear waves to an accuracy of about 1 % using the principle of the mode conversion (section 2.5.2) of compression waves on this lateral reflection (Fig. 5.3(a)). After mode conversion, some of the energy is propagated as shear waves at a velocity lower than that of the compression waves. Further mode conversion takes place with

* This may not be the case for digital flaw detection equipment (section 6.2.5) for which the image on the screen is manipulated so that the position of the peak indicates that of the reflecting surface.

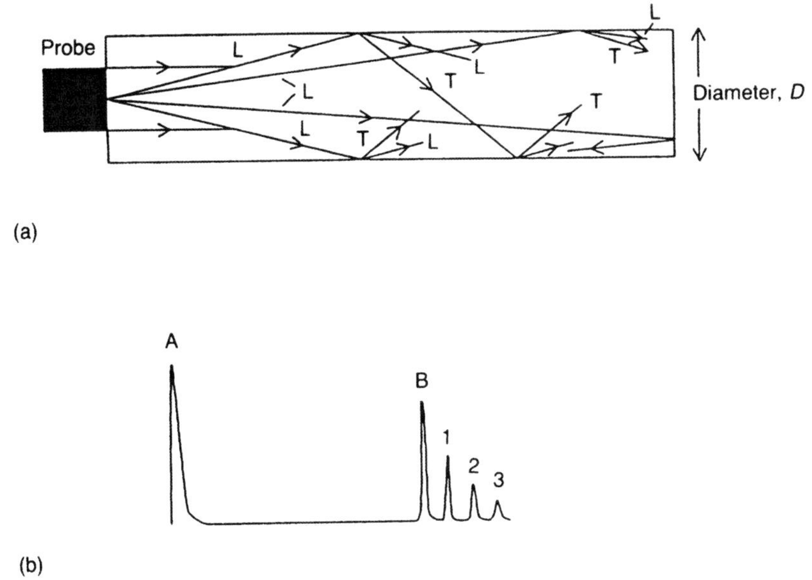

Fig. 5.3 (a) Mode conversion of compression waves at the lateral surface in the far field of a cylindrical rod of diameter, D; L and T indicate, respectively, compression and shear waves. (b) Displays of the back-wall echo B and echoes 1, 2 and 3 delayed successively by times δt; A = transmission peak.

subsequent lateral reflections, where some of the shear-wave energy is converted into compression-wave energy and vice versa. To produce this effect, the values of the length and diameters of the cylinder and the diameter of the probe should have values suitable to allow at least two mode conversions to take place: compression to shear waves and shear back to compression waves. Consequently, the indications on the oscilloscope screen include, in addition to the first back-wall echo, a number of equally spaced additional echoes of decaying amplitude (Fig. 5.2(b)). It can be shown (Appendix B) that the shear-wave speed, c_T, is given by:

$$c_T \approx c_L/(1 + a^2)^{1/2}, \tag{5.1}$$

where $a = c_L\delta t/D$, δt is the spacing between adjacent echoes and D the diameter of the cylinder.

5.3.3 Attenuation measurements(see also section 8.4)

The ultrasonic attenuation in a parallel-sided object is measured by comparing the heights of two back-wall echoes by means of the attenuator which is calibrated in decibels and can usually indicate levels to the nearest 1 or 2 dB. To avoid complications, the measurements should be conducted with peaks which are both **entirely** within either the far or near field. A correction of 6 dB for doubling of distance should be subtracted from the measured attenuation in the former case. For this reason, the orders of the observed peaks should bear a 2 to 1 ratio: the first and second order or the second and fourth, provided that the above conditions are met.

With the attenuator set for a convenient reading, the higher-order of the pair of echoes should be observed first and the sensitivity control adjusted until the tip of the peak just touches a horizontal line on the graticule in front of the oscilloscope screen. The attenuator setting is then adjusted until the peak of the lower-order echo touches this line. If an exact adjustment is not possible, the two consecutive settings of which one is above and the other below this line should be observed and the value obtained by interpolation. The difference, A, between the readings for the two observed echoes is that attenuation for the total path length between the corresponding reflections and the attenuation coefficient is equal to $A/2dn$, where n is the difference between the orders of the echoes and d the thickness of the sample. Thus for the first and second echo, $n = 1$.

To obtain accurate measurements of the attenuation coefficient by this method it is important that the transducer in the probe is correctly mounted and that all surfaces (those of the sample and the transducer), are accurately parallel to one another. Failure to do so results in the lowering and broadening of the peaks thus bringing about incorrect indications. Furthermore, it is important that the operating frequency and bandwidth are precisely defined for a quoted value. This is important, for example, for accurate estimations of grain size from considerations of Rayleigh scattering (section 3.3). However, for applications such as routine defect sizing or rough estimations of grain structure, this degree of stringency may not be necessary. It is also important to ensure that spurious indications are not present (section 5.3.4).

Attenuation measurements can be used for damage evaluation. The presence of a sufficient degree of damage in an object gives rise to attenuation of the back-wall echo with the pulse-echo method and, with the shadow method (section 5.4), of the transmitted signal. Prassianakis (1994) has reported the application of the pulse-echo method to observe the effect of widespread defects which, individually would be unlikely to be large enough to produce detectable echoes. However, the collective effect of these defects could give rise to scattering and thus an observable decrease in the height of the back-wall echo as a result of the resultant attenuation. the test specimens, in the form of aluminium bars, were progressively stressed until fracture occurred. During the stressing, the height of the back-wall echo was found to decrease with increasing rate so as to enable the evaluation of the progress of any material damage. To quantify the amount of material damage, Prassianakis considered the parameter $D = \delta S_0/\delta S$ where δS_0 is the cross-sectional area of the sample with respect to the direction of the waves and δS is the effective area presented by the distributed defects. Results were compared with those obtained by monitoring the degree of material damage by measuring electric currents passed through similar samples and good correlation was found. It was established that the ultrasonic method simpler and more rapid than the use of electric currents. Applications of attenuation measurements in polymers for defect evaluation are discussed in section 10.4.2.

Degrees of scattering can also be evaluated by observations of pulse echoes reflected back to the source by backscatter, which is discussed briefly in section 5.8.

5.3.4 Spurious indications

With the pulse-echo method, spurious peaks may appear on the screen of the oscilloscope as a result of the following:

- generation of a fresh pulse before its predecessor has died down;
- echoes in a delay medium (buffer); and
- lateral reflections giving rise to mode conversion.

The brightness of a signal can be controlled by increasing either the rate of emission of the electron beam in the oscilloscope or the pulse repetition frequency. To avoid undue wear on the oscilloscope tube the latter choice is to be preferred. However, for a material with a low attenuation coefficient, it can often happen that the next pulse is generated before the existing one has been fully attenuated thus giving the possibility of a 'ghost' echo being observed. Taking the example of a material in which the speed and attenuation coefficient are $5000\,m\,s^{-1}$ and $5\,dB\,m^{-1}$, the time taken for each pulse to reduce in amplitude by 100 dB is 4 ms, during which time they have travelled 20 m which corresponds to 10 back-wall echoes in an object 100 mm thick. For flaw detection purposes, the time base is expanded so that only the first of these echoes is observed. To avoid the appearance of any 'ghost' echo resulting from a higher order reflection of the previous pulse, either the pulse repetition frequency should be kept to below 250 Hz or the dynamic range of the instrument be reduced by a suitable amount by decreasing the sensitivity of the amplifier. However, these ghost echoes can usually be recognized easily.

Multiple echoes occurring in a buffer, such as a Perspex shoe or an immersion liquid, can be concealed by ensuring that the time delay here is greater than that for the first back-wall echo in the test object so that the first back-wall echo in the sample appears on the screen before

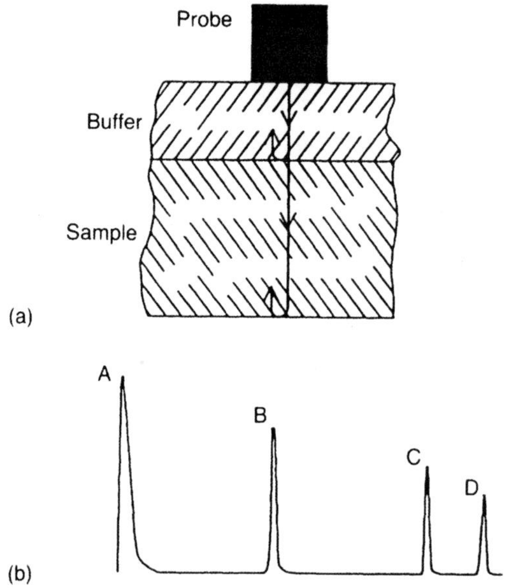

Fig. 5.4 (a) Insertion of delay medium (buffer) D between test sample S and probe P. (b) Resulting indications on the screen of the flaw detector; B = first front-wall echo, C = first back-wall echo and D the second front-wall echo, all from the sample.

the second one in the buffer (Fig. 5.4). For example, if a steel block (speed of sound $5900\,\mathrm{m\,s^{-1}}$) is immersed in water (speed $1500\,\mathrm{m\,s^{-1}}$), the ratio of the speeds is nearly 4 to 1. Provided that the thickness of the steel is less than four times the depth of its upper surface below the surface of the probe, no spurious echoes would be observed. Here the time base is delayed and expanded so that the echo of the front surface of the steel coincides with the zero of the graticule scale on the oscilloscope and that from the back surface appears at the right-hand extremity of the screen.

The phenomenon of lateral reflections undergoing mode conversion is discussed in section 5.3.2 with respect to the measurement of the speeds of shear waves in cylinders. In other circumstances, peaks appearing in this manner are sometimes a nuisance. For example, and accurate measurement of attenuation by observing multiple echoes would not be possible if some of the energy has been removed from the regular display pattern as a result of this phenomenon. This difficulty may be overcome by noting where the spurious peaks commence and by restricting measurements to earlier echoes. If, by doing this, one peak lies within the near field and the other within the far field, the use of a transducer with a larger diameter might overcome this difficulty. Trouble may occur when inspecting shafts, axles and other long objects in the axial direction for defects, especially when there are abrupt changes in cross-section. Identification of spurious peaks can usually be achieved either by testing a known defect-free sample and identifying the echo peaks from its geometry or by moving a wetted finger tip across the surface and noting any decrease in peak height as a result of sound transmitted into the finger at positions of lateral reflection.

When using transmitting and receiving probes located in separate housings, there is a possibility that surface waves, excited by the transmitter and picked up by the receiver, produce a spurious signal but the latter can be identified by moving one of the probes across the surface and observing any peak movements, but bear in mind that internal path-lengths can also changes as a result of this action. This type of signal is unlikely to appear when the transducers are mounted on buffers.

5.4 THE PULSE-TRANSMISSION (OR SHADOW) METHOD

With the pulse-transmission (or shadow) method, two probes, one transmitting and the other receiving are located in line with one another and opposite sides of the object being tested. The method consists of observing the received peak and relating it to a standard, such as an electrical output signal from the transmission probe. This method is often used for testing highly attenuating materials where the amplitude of any echo would be too low for any satisfactory evaluation. It has also been found useful for testing materials having uneven surfaces for which observed echoes are indistinct but transmitted pulses would still contain useful information. Here, coupling is achieved either with the samples and transducers all immersed in a liquid or by suitably applying jets of water (section 4.4.2). The presence of a defect gives rise to a reduction in the amplitude of the received signal by an amount which can be correlated to the characteristics of the defect.

The through-transmission method can be used to measure acoustic velocities and thicknesses of samples. For objects of suitable size and shape, transmitting and receiving probes are immersed in a water tank containing a thermometer and located opposite one another at a fixed distance apart. The peak corresponding to pulses received directly from the receiver is suitably positioned on the time base of the oscilloscope. On inserting a parallel-sided test object between them, there is a shift in the position of the peak which corresponds to the resulting change in transit time, which is measured. If c and c_W are the respective speeds of ultrasound in the test object and water, d the thickness of the sample and δt the change in transit time, we have:

$$\delta t = d(1/c_W - 1/c).$$

The speed of sound in water at the temperature indicated on the thermometer can be obtained from published tables (Kaye and Laby 1973).

5.5 THE PULSE-RESONANCE METHOD

The pulse-resonance method provides a convenient method of measuring thicknesses of plates and sheets. It can also be used for testing bonds. In one form of the pulse-resonance method, long pulses (of narrow frequency band) are propagated by a broad-band transducer (such as a flaw-detector probe), excited by a constant amplitude radio-frequency oscillator, through a delay medium to the test sample. Provided that the pulse width is equal to at least twice the thickness of the sample, standing waves can be set up in the latter (section 2.6). The sample then undergoes forced vibrations and transmits pulses which can be detected by the transmitting probe or a suitably located receiving probe, for example on the other side of the sample. The pulses are then displayed on the oscilloscope screen, in their unrectified form, giving the appearance of a train of waves. For ease of observation, the time-base is adjusted until the received signal covers most of the screen. On changing the input frequency in a continuous manner, the amplitude of the waves in the pulse increases to a maximum, when a state of half-wavelength resonance is reached, decreases to a minimum and then increases to a maximum at the next half-wavelength resonance for which:

$$f_r = nc/2l, \tag{5.1}$$

where l is the length (or thickness) of the test sample, c the speed of sound in it and n the harmonic order (section 2.6.5). The fundamental frequency f_1 is equal to the difference between two adjacent harmonics: f_{n-1} and f_n. l can then be calculated if c is known or vice versa. To improve the accuracy of measurement, changes δf in frequency between as many resonances as can be observed should be measured. Thus if 10 consecutive states of resonance can be obtained, we have $f_1 = \delta f/9$. An advantage of this method is that measurements can be made on samples remote from the probe, provided that adequate acoustic coupling can be maintained.

An alternative application of the pulse-resonance method is the introduction of wide-band pulses into a sample through a delay medium and frequency analysis of the received pulses. The observed frequencies are the resonance frequencies of the sample and the thickness can be determined by applying equation 5.1 as before. This technique has been used for testing bonds (section 10.7.6) and also for ultrasonic measurements at high temperatures of an object in an oven or a furnace (Bell 1968). For this purpose the transducer is maintained at a constant (room) temperature and coupled to one end of a long rod made from a refractory material. At the other end, located in the high temperature zone, there is an abrupt change in cross-section providing a final section having a length to give rise to resonance within the available range of frequency. From the value of the resonance frequency, the temperature can be determined if the device has been previously calibrated for known temperatures.

5.6 SCANNING AND IMAGING

To carry out flaw detection in a thorough manner, it is necessary for a complete scan of the surface of the test object to be made. To do this, the probe is moved in a systematic manner, e.g. along parallel lines, over one or more surfaces until the entire region in which defects are likely to be found has been swept with ultrasound. Sometimes it may be more convenient for the probe to be kept in a fixed position and the test object moved. To ensure a complete scan, it is important that the paths traced out by the centre of the probe are not spaced apart by an amount greater than the diameter of the radiating surface of the probe, otherwise some regions of the object may remain untested. Scanning can be performed either manually or automatically and, to avoid any variations of pressure of the probe on the object both the latter should be immersed in water or other suitable liquid with their surfaces maintained parallel with one another and equally spaced apart. With the through-transmission method, the two probes are maintained rigidly at a fixed distance apart. Four different scanning techniques, i.e. A-, B-, C-, and P-scans are discussed here. The B- and C-scan methods are restricted to use with normal probes whilst the P-scan technique has been developed for angle-probe applications. The more recently introduced D-scan, which involves moving two probes at a fixed distance apart, is described in section 8.2.3.

The **A-scan** is the most commonly used scanning method. Here an ultrasonic flaw detector is used in the conventional manner (section 5.2) and the indication on the oscilloscope screen monitored for all probe positions. With automatic scanning, the echo signals are fed to either a recorder or a computer in which data of the position and height of these signals are matched to those for the probe position, thus enabling defects to be readily located. Provision can be made for an alarm to be actuated either when a defect echo is located or when it has an amplitude exceeding a predetermined value, thus allowing, if necessary, for a further investigation. A-scanning can be performed with both compression and shear waves.

With the **B-scan**, using the conventional testing methods, a storage oscilloscope is used to retain the image during the duration of the scan. The time base is fed to the Y-plates of the oscilloscope and the output of a probe manipulator, which also indicates the position of the probe, is connected across the X-plates. The oscilloscope beam thus moves horizontally across the screen as the probe is moved in the x-direction. The output of the probe in its receiving mode is fed to a device which modulates the intensity of the beam to correspond to the value of this output: a bright spot indicates a high echo amplitude and a dull spot indicates a low one. Bright spots appear at the top of the screen to indicate the back-wall echo and also the front-wall echo with either an immersion technique or the use of a buffer. The presence of a defect thus gives rise to a series of spots of screen height corresponding to the defect depth and the brightness of these spots indicate its nature, since it affects the reflection coefficient. With a strong echo, there may be a corresponding decrease in the brightness of the corresponding part of the back-wall echo. The display appears on the screen of a storage oscilloscope or similar device and remains visible until it is cancelled. The effect of the scanning is to produce a series of adjacent spots which appears as a continuous line. Thus there appears two parallel continuous lines representing the front- and back-wall echoes and, if appropriate, other lines representing defects (Fig. 5.5). A single B-scan thus provides a visual image of the internal structure of a vertical section of the sample within a thickness equal to the diameter of the probe assuming that there is no far-field beam spread. The whole

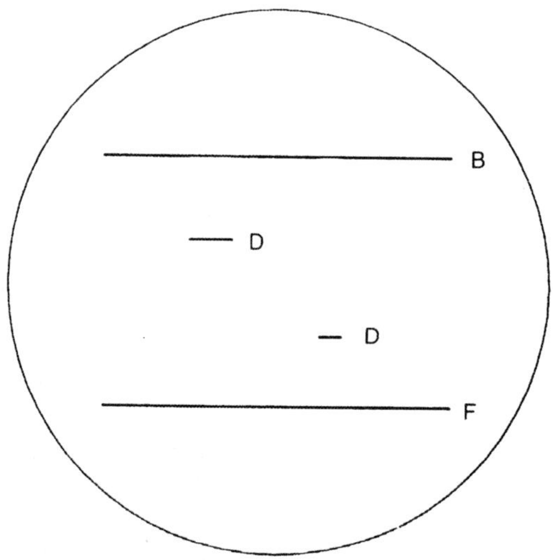

Fig. 5.5 Screen display for B-scan flaw detection; F = front-wall echo, B = back-wall echo and D = defect.

of the internal structure can be covered by making the appropriate number of individual scans. The far-field spread can be counteracted by the use of a suitable focused probe (section 2.10).

With **C-scanning**, which is now widely used for testing relatively thin samples such as sheets of steel or carbon-fibre composites, the probe is moved in two dimensions. The motion of the probe in one direction controls the voltage across the X-plates of the oscilloscope and that at right angles to this direction controls the voltage across the Y-plates. As before, the beam is modulated by the received ultrasonic pulses and an echo is indicated on the screen by a spot in a position corresponding to the point of reflection and with a brightness consistent with the echo amplitude. Although there is no longer any time base in the circuit, it is possible to estimate the depth of an echo by means of a gating device which can be manipulated to allow the detection of echoes at any desired depth, with the exclusion of those at other depths. With C-scanning, the trace on the oscilloscope screen takes on an appearance similar to that of a radiograph. When testing thicker samples, the depth of an echo can be evaluated by performing two scans along surfaces perpendicular to one another. Alternatively, the echo depth can be obtained approximately by repeating the scans with progressive gating so that different depths can be viewed in turn.

B- and C-scanning can be done with an ordinary flaw detector if either its output is fed into a computer for future recall, as indicated below, or the screen is shielded from ambient light and viewed by a camera containing a film exposed for the duration of the scan. The C-scan method can be adapted to chart recording for which the motion of the probe is linked to a pen or other recording device. A pen can be constructed to release inks either of different colours (or of varying monochrome levels) depending on the amplitude of the echo. A simple example of this method using different amounts of black (section 10.5.3) is the scanning of large sheets of fibre-reinforced plastics using the through-transmission method. Here, three ranges of amplitude are indicated, i.e. from up to 8 dB by black, 8 to 12 dB by grey and over 12 dB by white.

With the aid of a suitable computer program it is now possible to conduct A-, B- and C-scans simultaneously without the need of an oscilloscope (except to act as a monitor). The results of a required scan can thus be obtained in the form of either a screen or printed graphics display, or printed indications of position, size and location of any discontinuity in the object being tested.

B- and C-scanning cannot be applied directly to shear-wave testing because the echoes occur in positions laterally displaced from the probe and not beneath it. This difficulty can be overcome by the use of a projection scanning method, the **P-scan**, which is sometimes applied to the testing of welds. Information about the echo time delay, the probe angle, the speed of shear waves in the test object and the echo amplitude are fed into an electronic circuit. This gives rise to a modified B-scan display in which horizontally projected, instead of vertical, distances are displayed. An account of this technique is given by Krautkrämer (1990).

Further information on scanning techniques is given in section 7.5.

5.7 CONTINUOUS WAVE TECHNIQUES

5.7.1 Progressive wave methods

Continuous wave measurements make use of either progressive or stationary (standing) waves. With progressive waves, the ultrasound is passed through the test sample and picked up by a receiver; the phase and/or amplitude are compared with those of the transmitter. The use of this method is generally confined to testing highly attenuating materials, otherwise unwanted standing waves appear but, in some instances, this may be prevented by applying some damping material to the end surface or by suitably shaping that surface to inhibit reflections.

One application of continuous progressive waves is the measurement of compression wave speed and attenuation in long rods having diameters small compared with the wavelength, a condition which requires the use of relatively low frequencies, typically below 100 kHz. The transmitting probe is located at one end (Fig. 5.6) and, if necessary the other end is treated to inhibit reflection. The receiving transducer is located on the side of the rod where it can be made to slide along the

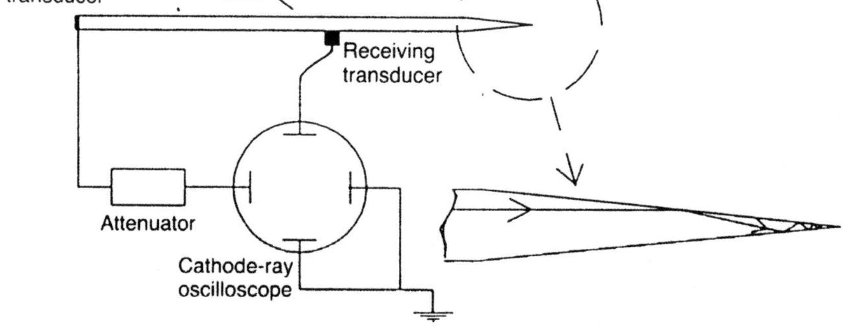

Fig. 5.6 Continuous-wave method for measuring the speed of compression waves in a rod. Inset shows decaying multiple lateral reflections at the tapered end of the sample.

surface so as to pick up any lateral components of the waves travelling in the axial direction (section 3.2). The input of the transmitter and the output of the receiver are connected, respectively, to the X- and Y-plates of a cathode-ray oscilloscope. The voltage output appears as an ellipse, the shape of which depends on the amplitude of the received waves and the phase angle. By adjusting an attenuator located between the transmitter and the oscilloscope, the amplitudes of the two components can be equalized, thus facilitating observation. The distance moved by the receiver corresponding to a phase change of 360°, as indicated by the change in shape of the ellipse (Fig. 3.1)), is equal to one wavelength, λ, and, from a knowledge of the applied frequency, f, the speed of sound $c = \lambda f$ can be calculated. The attenuation coefficient can be determined by comparing the amplitudes of the received signals at different distances from the source. This is most easily done by replacing the output of the transmitter to the X-plates with that of a time base so that a sine wave display of constant amplitude appears. The amplitude ratio for different positions can then be compared with the aid of an attenuator calibrated in decibels: if a displacement of x m is required for a change of n dB in the attenuator setting, the attenuation coefficient is simply $n/x\,\mathrm{dB\,m^{-1}}$. It should be noted that when the rod has a diameter small compared with the wavelength, for which the ultrasonic speed is given by equation 3.2, plane-wave conditions apply for **all** distances from the source.

5.7.2 Stationary wave methods

Except for highly attenuating materials, the propagation of continuous waves in a body having a finite length leads to the setting up of stationary waves. Methods employing stationary waves can be used to determine values of wavelength (and, hence, speed of sound or thickness) and attenuation coefficient. This is done by obtaining conditions of resonance by varying either frequency, for both solids and fluids, or distance, for fluids only, until the measured distance is equal to an integral number of half-wavelengths or an odd number of quarter-wavelengths (section 2.6).

The resonance technique is sometimes used for thickness gauging. Here, a wide-band piezoelectric probe is placed in contact with the test object, with suitable coupling, and excited by means of a variable radio-frequency oscillator (section 5.5). Half-wavelength resonances in the material are indicated when the detected signal is a maximum, as shown by means of either a meter, an audible sound or a peak on the screen of an oscilloscope. In the latter instance, the exciting frequency is swept through a given range, such as 2–6 MHz, and peaks indicating resonances appear; the time base of the oscilloscope is subjected to a Fourier transformation (equations 2.60(a) and (b)) so that it indicates frequency. The distance of separation of two adjacent peaks is then equal to the fundamental resonance frequency, f_1, and the thickness of the material is then equal to $c/2f_1$. Graticules calibrated for the material of the sample under test may be placed in front of the screen and the thickness read directly. It should be noted that this device has been largely superseded by the simpler and more convenient digital thickness meter based on the pulse-echo technique (section 6.3).

To measure attenuation in the lower kilohertz range in a rod or bar, the **damping capacity method** may be used. The test sample is supported by lightly clamping at a nodal position, usually at the centre, and either allowed to vibrate freely at resonance after shock excitation or made to undergo forced vibrations over a swept frequency range about the resonance frequency. Fig. 4.7(b) illustrates a system in which a sample in the form of a rectangular-sectioned bar, nodally mounted, with electrostatic transducers at each end, one acting as a transmitter and the other as a receiver. A similar system can be employed, using electromagnetic transducers (section 4.5.1). The fundamental resonance frequency is determined by the length of the sample equal to one half-wavelength. Thus for an aluminium rod 64 mm long, in which the speed of compression waves is equal to 5100 m s^{-1}, the fundamental resonance frequency is approximately 40 kHz.

With the shock excitation method, the transducer performs free damped oscillations in the same way as an ultrasonic probe used for flaw detection: with a high direct signal. If this is done at regular intervals by means of a trigger synchronized with the time base of an oscilloscope, a display similar to that seen in Fig. 2.2(b) is observed. The logarithmic decrement, δ, and hence the 'Q' factor (equations 2.4 and 2.9) can be obtained from this display. If the signal is fed to the oscilloscope by means of a logarithmic amplifier, the exponential decay of the curve becomes linear and the value of δ is obtained from the gradient of this curve. Alternatively, the signals from the receiving transducer can be fed to a digital counter which registers the number of vibrations performed while the amplitude has decayed to a fraction of 1/e (−8.686 dB) of its original value. From equation 2.4 we seen that $\delta = 1/n$.

With the swept frequency method, the sample is excited over a continuous range of frequencies with a constant input voltage amplitude applied to the transducer and the variation of current amplitude with frequency observed. When the frequency is swept at regular intervals of time in synchronization with the time base of the oscilloscope, a curve similar to that illustrated in Fig. 2.3 is displayed on the screen, from which the 'Q' factor and thus δ can be determined. Thus ordinates of this curve are proportional to the transducer current and, hence, to the velocity amplitude as obtained by measuring the potential difference across a resistor in series with the transducer.

5.8 THE ULTRASONIC GONIOMETER

The ultrasonic goniometer (Mayer 1965) is based on the principles of reflection and transmission at oblique incidence at a plane boundary (section 2.5.2). It can be used to measure the speeds of compression, transverse and Rayleigh waves irrespective of the shape of the test object provided that it has at least one plane surface. The method is used to its best advantage with the pulse-echo technique using two probes, one transmitting and the other receiving and each of which are mounted on rotating arms, the angular positions of which can be accurately determined with the use of verniers on a circular scale. The complete arrangement is immersed in a water tank. The test object is positioned in such a way that the axes of the probes intersect at its surface, which is perpendicular to the plane containing these axes (Fig. 5.7). The mechanism is designed in such a way that both arms can be rotated simultaneously in such a way that the axes make equal angles with the normal to the point of incidence, for all angles of incidence, so that the reflected compression waves are picked up by the receiving probe.

It can be shown that there is an abrupt change in the amplitude of the reflected compression waves when a critical angle is obtained with incidence at a liquid–solid surface (Fig. 2.9). By observations of such changes, the speeds of compression, shear and Rayleigh waves can be determined from the equations 2.49(a), (b) and (c), respectively, if the speed of compression waves in water is known.

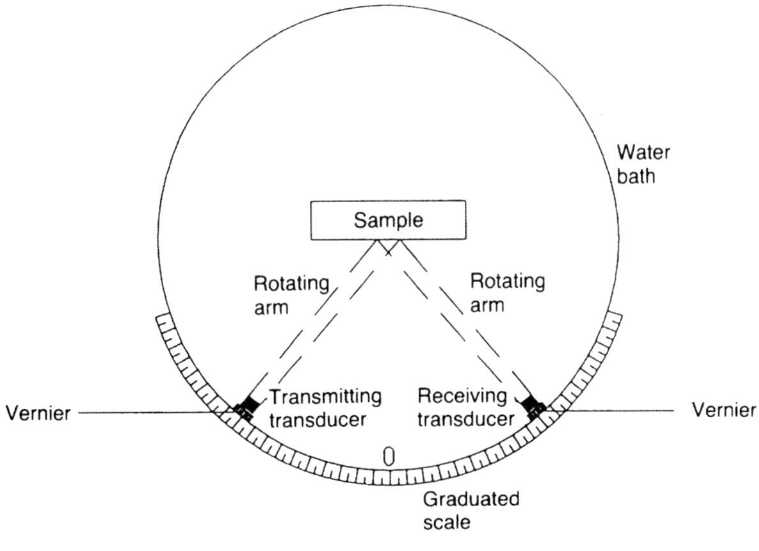

Fig. 5.7 Basic design of ultrasonic goniometer.

To achieve a high degree of accuracy, the ultrasonic beam should be parallel, for which it is necessary for the total acoustic path length to remain within the near field. Furthermore, it is essential to ascertain that the axes of the beam and probe coincide for both transmitting and receiving transducers.

The attainment of near-field conditions presents little difficulty at megahertz frequencies because of the lower acoustic speed for water compared with that typical for metals. For example, with the use of a 25 mm diameter probe at a frequency of 5 MHz, the wavelength in water is about 0.3 mm and the corresponding length of the near field is 500 mm, thus allowing for a maximum probe-to-object distance of 250 mm. This method is very advantageous for highly absorbent materials, such as a number of commonly-used plastics. It provides also a convenient and reliable means of measuring the speeds of surface waves (section 5.10).

An indirect method of measuring attenuation involves the observation of pulse echoes reflected back to the source by **backscatter**: that portion of scattering which takes place in the reverse direction (section 3.3.1). This technique has the advantage of allowing measurements to take place at higher frequencies because it is the increase in the amount of scatter with frequency which is measured and not the back-wall echo level, which requires a return transit through the object under test.

With the use of the goniometer, the angle of incidence can be set to a value allowing mode conversion to shear waves with optimum efficiency (equation 2.50(a)). The advantage of using shear waves is that the value of K in equation 3.9(b) increases with degree of scattering at a greater rate because of the lower speed of sound compared with that for compression waves. With the pulse-echo method, the degree of backscatter can be assessed from the fine structure of the echo pattern which arises from reflection at the individual sources of scatter, such as grains and inclusions.

Applications of ultrasonic backscatter measurements include the monitoring of stressed samples (Bridge and bin Ismail 1984), which is claimed to provide a sevenfold degree of sensitivity over results obtained from velocity measurements, and examining degrees of surface roughness at frequencies of 5 and 13.4 MHz (Bridge and Bollini 1987) by mode converting the incident waves to surface waves at the boundaries of the samples. Reports are available from Leviston and Bridge (1988) on the evaluation of the sub-surface microstructure of quenched and hardened carbon steel and from Bridge and bin Saffrey (1988) on the monitoring of annealing temperatures. Schlengermann and Schmitz (1994) describe the measurement at a frequency of 20 MHz of the hardening depth in a steel sample by the use of a basic immersion technique. The probe was set at the angle of incidence (17°) which allowed a maximum intensity for the mode-converted beam of transverse waves (at an angle of refraction of 40°) (section 2.5.2).

5.9 IMPEDANCE MEASUREMENTS

In section 2.3.4 the specific acoustic impedance, Z_a, of a material is defined as the ratio of acoustic pressure to particle velocity and is said to be a complex quantity having a resistive component, R_a, equal to the product of the speed of sound, c, and the density, ρ, and a reactive component, X_a. For plane progressive waves, X_a is generally accepted as being equal to zero but this is not the case for a material having a measurable absorption coefficient, α, for which X_a is $-\rho c\alpha/k$ (Appendix C). The tangent of the phase angle ϕ between X_a and R_a is thus equal to $-\alpha/k$.

The technique can be applied to the testing of plastics, rubber and other highly attenuating materials. With one method (Hueter and Bolt 1955), a resonating transducer is placed in contact with a standard material with a low attenuation coefficient and having known values of density, ρ_0, and of speed of sound, c_0, i.e. $Z \approx \rho_0 c_0$. It is then tuned to its fundamental resonance frequency and the amplitude noted. On then placing the transducer in contact with the test material a change in both amplitude and frequency takes place; the transducer is then retuned and reset in amplitude to its conditions when in contact with the standard material. From the change in frequency and amplitude, the components of the specific acoustic impedance can be determined (see also section 10.7.5).

5.10 SURFACE AND LAMB WAVE MEASUREMENTS

Surface (Rayleigh) waves (section 2.5.2) can be propagated in solids in the same way as with shear waves by using an angle probe (section 4.4) provided the probe angle is sufficiently large to allow mode conversion. Alternatively, they can be introduced by a compression wave probe into the solid, when both are immersed in a liquid, provided that the probe is inclined to the solid at an appropriate angle. This latter method has an advantage that the angle of incidence can easily be adjusted to an optimum value. Figure 5.8 illustrates the detection of a surface crack in a solid sample using the pulse-echo method and indications on the oscilloscope screen of echoes from the crack and from the edge of the sample.

As indicated in section 3.5, Lamb waves can be propagated along thin sheets of material having thicknesses of about a millimetre or less and this may be done by introducing surface waves in the manner

(i)

(ii)

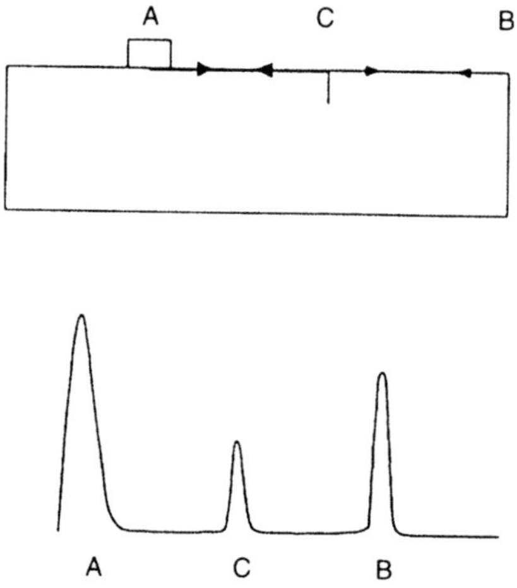

Fig. 5.8 (i) Generation of surface waves in a solid sample by the pulse echo method with the use of a large angle probe A. C indicates crack and B the first edge at which reflection takes place. (ii) corresponding display on the screen of an ultrasonic flaw detector.

indicated above. Because the order of the excited mode depends on the frequency and the thickness of the sheet, the use of a wide frequency band probe ensures the excitation of a mode appropriate to the thickness of the sheet, for a given angle of incidence.

Defects in a thin sheet can be detected by Lamb waves with the single-probe pulse-echo method by observing echoes characteristic of discontinuities in the same way as with surface waves. The pulse-echo method, using two probes, has been found suitable for detecting subsurface laminations in a solid and also lack of bonding of a thin sheet on a thicker substrate (section 10.7.4). Alternatively, pulsed surface waves can be propagated using a single broad frequency band angle probe so that when reaching a thin plate of the material formed between its upper surface and a laminar defect, the plate is excited to resonance at one or more of the Lamb-wave modes and the pulses generated by these excitations are detected by the probe.

5.11 ACOUSTIC EMISSION TESTING

The technique of acoustic emission is usually applied to what is called **condition monitoring**, the continuous testing of a structure or component whilst in service, rather than the more common practice of testing only at regular intervals. Because it is a highly specialized and somewhat complicated subject, a large amount of space would be required to give acoustic emission testing adequate treatment in this book. For this reason, only a short summary is provided here and the reader is recommended to consult McIntire (1987) for a comprehensive treatment of the subject.

When a solid is subjected to a stress beyond the yield point, it undergoes plastic deformation and, eventually, fracture takes place giving rise to crack growth. Part of the energy of fracture causes propagation, through the material, of acoustic waves covering a wide band extending from audible to high ultrasonic frequencies. These waves can be received by a highly sensitive transducer and the formation and growth of cracks in highly-stressed regions can be monitored, i.e. a crack can be detected by acoustic emission at the moment at which it is initiated. By placing a number of transducers in suitable positions, a source of emission can be located by the means of a somewhat complicated triangulation technique by determining the time differences between arrivals of a given emission at the transducers (Scruby 1985).

The acoustic emission method has been applied to the detection of incipient cracks, for example, in pressure vessels, aircraft and chemical processing plants which are subject to fatigue when in operation. It has also been used for the determination of the onset of stress corrosion cracking and for monitoring any crack formation during welding and crack growths during fracture mechanics investigations; in the latter instance, the test pieces are subjected to progressively increasing stresses. The method has been found to be particularly advantageous for the detection of leaks in pipe lines and other vessels subjected to high-pressure (Williams 1980). Because of the very low intensities of the emissions, it is essential to use a highly sensitive piezoelectric transducers, such as one of the PZT variety, in conjunction with a low noise preamplifier. Acoustic emission generally includes compression, shear and Rayleigh wave propagation and the choice of receiving probe is related to the type of waves providing the most suitable shape of signal: a short one or one of high amplitude. The choices of frequency and bandwidth are determined by the nature of the object being tested and the type of wave received; account should be taken of the fact that the attenuation coefficient usually increases with frequency in a manner depending on the nature of the waves. It is essential that the effects of any extraneous sound and ultrasound, especially within the frequency range under consideration, be eliminated.

A simple application of acoustic emission occurs in fracture mechanics investigations (Arrington 1987). This is the monitoring of the growth of a crack initiated by a saw-cut in a metal sample, subjected to a controlled tension, by placing it in a stress machine, and noting the relationship between the acoustic pulse emission rate and the length of crack. Another (Jiles 1988) is the study of the Barkhausen effect arising with the magnetization and demagnetization of a ferromagnetic metal, such as steel, where changes in magnetic flux density with magnetic field, which take place in a discontinuous manner, are accompanied by acoustic emission as a result of the motion of the magnetic domains in the material. The foregoing are simple and straightforward applications which need only one transducer. Complications, with consequent high financial costs, arise with larger scale location and monitoring of defects for which a number of strategically located transducers together with microprocessing equipment are required.

Further information on this subject, with special reference to the testing of fibre-reinforced composites, is given by Arrington (1987). The reader is also recommended to a short review paper by Raj and Jha (1994), which summarizes more general applications, and to a report by Rogers (1994) dealing with the sizing of fatigue cracks in offshore structures using acoustic emission.

REFERENCES

Arrington, M. (1987) Chapter 2 in *Non-Destructive Testing of Fibre-Reinforced Plastic Composites*, Vol 1 (ed J. Summerscales) Elsevier, London, pp. 22–63.

Bell, L.W.F. (1968) *Ultrasonics*, **6**, 11–14.

Bridge, B. and bin Ismail, M.P. (1984) *Nondestr Test Commun*, **1**, 155–64.

Bridge, B. and bin Saffrey, H.J. (1988) *Br J NDT*, **30**, 392–7.

Bridge, B. and Bollini, G.J. (1988) *Br J NDT*, **30**, 229–38.

Jiles, D.C. (1988) *NDT Internat*, **21**, 311–19.

Hueter, T.F. and Bolt, R.H. (1955) *Sonics*. Wiley, New York, pp. 354–68.

Kaye, G.W.C. and Laby, T.H. (1973) *Tables of Physical and Chemical Constants*, 14th edn. Longmans Green, London.

Krautkrämer, J. and Krautkrämer, H. (1990) *Ultrasonic Testing of Materials*, 4th edn. Springer-Verlag, Berlin, pp. 251–5.

Leviston, D. and Bridge, B. (1988) *NDT Internat*, **21**, 17–25.

Mayer, W.G. (1965) *Ultrasonics*, **3**, 62–8.

McIntire, P. (ed) (1987) *Nondestructive Testing Handbook*, 2nd edn, Vol 5, *Acoustic Emission Testing*. American Society for Nondestructive Testing, Columbus, OH.

Prassianakis, I.N. (1994) *Europ J NDT*, **3**, 93–6.

Raj, B. and Jha, B.B. (1994) *Br J NDT*, **36**, 16–64.

Rogers, L.M. (1994) *Insight*, **36**, 661–5.

Schlengermann, U. and Schmitz, G. (1994) *Experiences in the Determination of the Hardening Depth using Ultrasonic Backscatter*, Krautkrämer GmbH & Co, Postfach 1363, D-50330 Hürth, Germany, Publication ref. SD 284 (1/94).

Scruby, C.H. (1985) *Research Techniques in Nondestructive Testing*, Vol 8 (ed R.S. Sharpe). Academic Press, London, pp. 141–210.

Williams, R.V. (1980) *Acoustic Emission*. Adam Hilger, Bristol, pp. 66–8.

CHAPTER 6

Ultrasonic testing equipment

6.1 INTRODUCTION

The general principles of the design and uses of ultrasonic equipment are discussed in Chapter 5. In this Chapter the more commonly used items are dealt with in greater detail with stress laid on the latest developments.

In the 1950s when ultrasonic testing sets became generally available, they were invariably powered by mains electricity, contained vacuum tube (valve) amplifiers and were somewhat bulky. The screen displays were inclined to drift during use as a result of self-generated heat and cooling fans were often fitted to counteract this. The late 1960s saw the introduction of transistors in amplifier circuits, the attenuator calibrated in decibels and the more sophisticated multi-turn delay control to provide the shift of the cathode ray display in the x-direction. These devices vastly improved operating techniques and enabled greater ease in the uses of twin-transducer (usually called twin-crystal) and angle probes.

At the same time, advances were made in the miniaturization of equipment and its consequent portability as with, for example, pocket-sized ultrasonic thickness meters. Improvements in circuitry and clearer cathode-ray screen presentations followed and, by the late 1980s, rapid developments in microchips and computer technology had taken place. It appears that improvements in the design of ultrasonic testing equipment and its various accessories will continue at an alarming rate.

The various types of ultrasonic testing instruments are discussed in Chapter 5 and the principal ones are:

- high-frequency pulse operating equipment, generally used in the megahertz frequency range;
- low-frequency pulse operating equipment, used in the kilohertz frequency range; and
- damping capacity instruments.

Their characteristics are discussed in the following sections of this Chapter.

6.2 HIGH-FREQUENCY PULSE OPERATING EQUIPMENT

6.2.1 General considerations

The most commonly used ultrasonic testing instrument is the pulse-echo apparatus (section 5.3), usually known as the **ultrasonic flaw detector**. The pulse echo method is preferred where possible to the pulse-transmission technique (section 5.4) because of the greater amount of information received and the requirement for only one probe.

The output display consists of a cathode-ray tube or similar visual device (section 6.2.2) which is normally in the form of an A-scan: a 'real time' indication of the time delay, height and structure of the received echoes. Other scans, the B-, C-, D- and P-scans provide images of the acoustic field, as a result of reflections from defects and boundaries. The images are built up during a scan and may be displayed either on the screen of a storage oscilloscope or on a chart with the aid of a pen recorder. Details of the various methods of scanning are given in sections 5.6 and 8.2.3.

6.2.2 Components

Section 5.3.1 provides a summary of the various components of an ultrasonic pulse-echo instrument. It is the intention here to expand on the information provided in that section, with special reference to improvements in design and performances of the output display, pulse rate, amplification and noise suppression (or rejection).

Output display

Figure 5.1 provides a simplified block diagram of the pulse-echo instrument; this illustrates the use of a cathode-ray oscilloscope as the output device but the latest equipment replaces this with either a digitized or an electroluminescent display.

With the conventional cathode-ray tube the electron beam is deflected by electrostatic charges on the X- and Y-plates. These tubes are expensive, complex in design and very bulky. Recently, television receiver tubes (in which the electron beam is deflected by electromagnetic instead of electrostatic fields) have been introduced to display the output of the instrument. They are much cheaper, less complex in design and have considerably less depth than the original cathode-ray tubes. The outputs are digitized so that the display takes the form of pixels in common with domestic television sets. They have the advantage that a large-screen monitor can be connected to the equipment for display to an audience in a class room or lecture hall. A disadvantage, however, is that the display can be distorted if operation takes place in the vicinity of a strong magnetic field.

The latest form of display is an electroluminescent, liquid crystal (LCD), screen with solid state circuitry. This has virtually no depth but the visible area is large. It operates at a low voltage, so no high voltage supply is necessary; is very portable and takes up little room. For this reason, its use has become widespread. It has a variable time base which can provide corresponding ranges of from 10 mm to 10 m. The voltage across the X-plates is varied in such a way that the electron beam accelerates to a constant predetermined speed on reaching the zero position and remains at this speed until the final scale division, after which it rapidly comes to rest; the beam flies back to its original position and the process is repeated. Assuming that the beam takes 10 μs to move at constant speed across a time-base scale having 10 divisions, 1 division represents a time of 1 μs.

Pulse rate

In section 5.3.1 it is pointed out that if the value of the pulse repetition frequency (the pulse rate) is sufficiently high for a new pulse to be generated before its predecessor has died down below the noise level, then 'phantom echoes' appear. This is normally prevented by reducing the pulse rate but can be achieved automatically by the pulse-rate control being incorporated with the range control (section 5.3.1), with the result that the pulse rate is slowed down when the range is increased. With immersion testing, there is often a maximum rate of 1 kHz.

Pulse transmission

The pulse generator emits at intervals governed by the pulse repetition control, sometimes called a 'clock', a direct voltage of instantaneous duration having a value of from several hundred to a thousand volts, in accordance with the setting of the pulse energy control. The latter should be kept at a sufficiently low level to prevent overloading, with the consequent effect of 'noisy' echoes. For maximum power transfer, the impedances of the probe and/or connecting cable should be properly matched to that of the transmitter; the value of the impedance of the probe is very much dependent on the characteristic impedance of the transducer. An auto-transformer is usually installed inside the probe housing to assist in impedance matching.

Amplification

The provision of an amplifier in the receiving part of the circuit is essential in view of the fact that the voltage amplitude of the pulse received by the probe is reduced from an initial value of between 100 and 1000 V to one between 0.0001 and 1 V. For readable signals to be received by the oscilloscope, an input of several hundred volts is required and it is thus necessary to amplify the received pulses using either a linear or logarithmic amplifier. The former is limited to comparatively low values of amplification, as high values would give rise to distortion of the peaks. The amplifier is provided with a calibrated attenuator (or gain control) which adjusts the sizes of the echo signals on the screen. It is possible to alter, intentionally, the characteristics in steps, so that the echo signals appearing on the screen are all of equal height and so to obtain a 'swept gain effect'.

With the logarithmic amplifier, the amplitude is proportional to the logarithm of the output voltage of the probe and is measured in decibels (section 2.3.6). This amplifier provides a greater dynamic range, e.g. 100 dB compared with 34 dB with a linear amplifier. The logarithmic amplifier is not popular with the operators, probably because when using the 6 dB and 20 dB drop techniques (section 8.3), the changes do not look the same as those obtained with a linear amplifier.

Noise suppression (or rejection)

The noise suppression control is designed to prevent interference between noise in the time-base signal and small flaw echo signals. Its operation, however, can affect the linearity of amplification (or gain). At one time, noise suppression was achieved by lowering the echo display by a given amount below the time-base line; this had the unfortunate effect of lowering all the echo signal heights, both large and small, by equal amounts. In extreme cases, if the degree of suppression is too high, important flaw echo signals might disappear resulting in large flaws being undetected. As a result, many working procedures for tests ban the use of the suppression control. More recent equipment uses filtering devices to remove noise and, although they do not affect the heights or real echo signals, the latter become thinner and hence difficult to see. Another method of suppression is to apply an inverse feed, signal averaging, which reduces the random noise without affecting the real echoes.

6.2.3 Controls

The principal controls used in the operation of pulsed ultrasonic testing equipment are:

- the time-base control;
- the zero shift (or delay) control;
- the on/off control; and
- the gain control.

The aforementioned controls are fundamental to ultrasonic testing and are adequate for most basic operations. Those regarded as being optional include:

- the suppression/rejection control;
- the gate monitor;
- the pulse energy control;
- the frequency control;
- the swept gain control;
- pulse-rate control; and
- rectification control.

Time-base control

The time base is usually adjustable over a wide range, equivalent to 10 mm to 10 m and, although these ranges are continuously variable for any value within these limits, it is prudent to choose a range having a value which is readily divisible by 10 so as to be compatible with a zero to 10 scale. A knowledge of the speed of ultrasound in the material under test is of prime importance and settings of the range (the distance from the probe) are only meaningful if they can be related to the material. The range controls usually allow for both coarse and fine tuning and can be locked once the range has been calibrated.

Zero shift (or delay) control

A zero shift, or delay, control is essential to enable the zero position of the probe to be accurately set for calibration of the time base. It was originally designed to set the zero position of a signal on the oscilloscope screen by moving the time base in the appropriate direction. However, the amount of movement was found to be insufficient for many purposes, e.g. to show the delay produced by a buffer fitted to a probe, and the screen display would thus be incomplete. The use of a multi-turn delay control, typically a ten-turn potentiometer, allows a full-screen display of the available signals irrespective of any time delay with, at the same time, the coincidence of the near-surface echo and the zero position of the time base.

The time-base control can also be used to select any required short-distance range such as that from 0 to 10 mm and then to rotate the control so as to enable any range of this length, e.g. from 25 to 35 mm, to be displayed. In this way, required sections of a test sample can be studied in more detail; this procedure is called 'zooming'.

On/off control

The on/off control often incorporates the selector switch for single and twin crystal probe operations and, for this reason, it is easy to overlook the need for resetting this switch when changing one of these probe types for another; great care should thus be taken in this respect.

Gain control

The operation of the gain control, or attenuator, ensures a sufficient degree of sensitivity for a correct test procedure. It often happens that the setting of the gain control is laid down in a test specification. The use of gain controls for flaw sizing is described in section 8.1.2.

Suppression/rejection control

The suppression/rejection conrol, which is described in section 6.2.2, may be used during thickness gauging when it can be rotated by a sufficiently small amount to remove any noise and thus to obtained a sharper signal. Its use is generally not permitted with flaw detection in case any defects may be overlooked.

Monitor gates

Monitor gates are devices which can be used to monitor events occurring during a test. When the amplitude of the signal exceeds the threshold height of a gating device, an electrical impulse actuates an alarm system. The amplitude of this impulse is usually small but, if it is necessary to operate switches or solenoids to provide, for example, an audible signal, the use of a relay is necessary.

Monitors provide an invaluable aid to the operation of automatic systems and one has the choice of employing either single or multiple gates. They can be used in the following ways:

- to record the number of received signals within the gate;
- to act as a go/no-go actuator, for which the same signal output is actuated irrespective of the amplitude of the signal; or
- to provide an output proportional to the size of the signal within the gate.

A monitor gate has a function known as a **threshold**, which can be controlled to a desired level, e.g. 20% full-screen height, so as to indicate by means of an alarm that the flaw size is at an unacceptable level. Originally, the appearance of a gate on the output screen took the form of a square pulse resting on the time-base line but, with more recently developed equipment, the form has changed to a 'floating horizontal bar' at the threshold level (Fig. 6.1). Further controls include one to place the left edge of a gate on the time base and another to adjust the width of the gate by an amount, for example, to cover the section of the time base which is of interest. In some instances, a gate

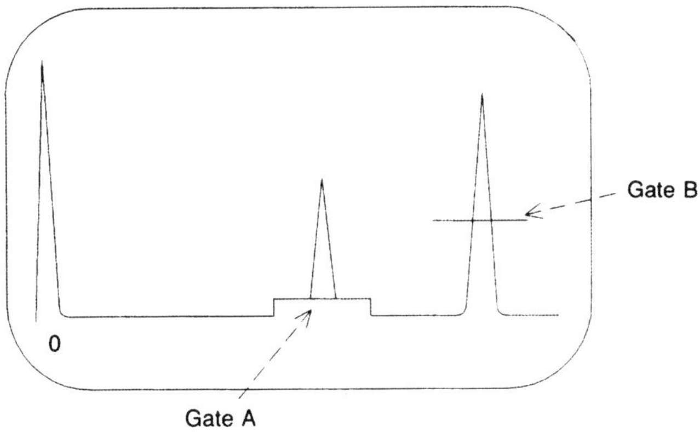

Fig. 6.1 Presentation on the screen of older flaw detectors of square pulse gate A and floating bar gate B.

can be made to operate in either a positive or negative manner so that, when set for positive operation, an echo rising from the time base, as a result of a flaw, triggers the alarm. When set for negative operation, the gate can monitor the back-wall echo and if the latter is either weak or non-existent as a result, for example, of a poor couplant, an alarm sounds so as to indicate an unsatisfactory test condition. However, if the back-wall echo signal has an amplitude greater than 80% of full-screen height, no alarm is sounded. Digitized ultrasonic flaw detectors use monitor gates to trigger measuring devices so as to provide a numerical display of further information such as distance and thickness.

Pulse energy control

The pulse energy control governs the output from the pulse generator to the probe. It can be set to a number of levels, usually five, for which the signal is increased in steps, e.g. from 100 to 1000 V. The setting is chosen to provide the minimum amount of noise.

Frequency control

Frequency controls are fitted to instruments having tuned amplifiers and the frequency of the latter is adjusted to match the frequency of the probe. The frequencies which are adopted are either 10, 5, 2.5 and 1.5 MHz or 12, 6, 4, 2 and 1 MHz but, at the time of writing, an international agreement has not been reached as to which of these patterns should be adopted. However, because the frequency bands for the tuned amplifiers and the probes are both broad, there should be sufficient overlapping to prevent any problem arising if the indicated probe frequency does not exactly match that of the tuned amplifier.

Swept gain control

The swept gain control, or time-corrected gain (TCG), is a device that amplifies received signals by amounts increasing with path lengths. Thus the multiple echo pattern (Fig. 5.2) caused by successive reflections at the parallel boundaries of a solid object of pulsed ultrasound, at right angles to the boundaries, indicates progressively decreasing amplitudes due to attenuation in the object. The effect of the swept gain control causes all of the displayed echo signals to have the same amplitude, thus simplify defect sizing.

Pulse-rate control

The pulse rate control is discussed in section 6.2.2.

Rectification control

The rectification control, which is often omitted, is used either to rectify or unrectify the displayed signals. The rectified display takes the form of either half-wave or full-wave mode with an option for either positive or negative phase for specialized applications. The unrectified display is used for examining wave forms and also for time of flight diffraction applications (section 8.2.3).

Interface trigger

The interface trigger is used when there is a variable delay path between the probe and test specimen as would occur, for example, with immersion techniques and when testing a thick coat of paint. It enables the user to 'lock on' to the interface between the delay path and the specimen, which serves as the position from which readings are made; the thickness readings are no longer affected if the delay path varies.

Ultrasonic flaw detectors

There is a wide variety of ultrasonic flaw detectors which ranges from small portable instruments for field work etc. to large mains operated assemblies used typically in laboratories and production lines. It is important that the user purchases a model which fulfils his/her needs sufficiently and not be seduced into purchasing sophisticated and expensive equipment containing seemingly attractive devices which serve no useful purposes.

Figure 6.2 depicts the analogue display on the screen of an ultrasonic flaw detector containing an electrostatic type of cathode-ray tube. Figure 6.3 shows an instrument having a digitized tube with electromagnetic focusing. Figure 6.4 depicts the more recently developed high-contrast electroluminescent display (section 6.2.2) and its presentation takes on a similar appearance to that of the digitized tube depicted in Fig. 6.3. It has advantages of low voltage and power operation and compactness, in that a thin flat panel replaces the more cumbersome tube. However, the smoothest display is that provided by the more conventional electrostatically operated tube. It should be pointed out that a certain amount of time is taken up in becoming familiar with the digitized instruments.

The operation of digitized display instruments is based on conversion of the continuously variable output of conventional analogue signals to digital form by sampling them at regular intervals. If these signals are

Fig. 6.2 Analogue presentation on the screen of an ultrasonic flaw detector having a cathode-ray tube with electrostatic focusing. Reproduced by courtesy of Krautkrämer GmbH & Co.

Fig. 6.3 Presentation on the screen of an ultrasonic flaw detector having a digitized television tube with electromagnetic focusing. Reproduced by courtesy of Krautkrämer GmbH & Co.

sufficiently short, the display bears a close resemblance to that obtained from the analogue signals. With this process, the original analogue signals are reduced to a series of digits which are converted into binary numbers, or bits. Once the signals are in digital form they are suitable for processing by means of a computer and can then be conveniently stored (Highmore and Short, 1988).

Digitized ultrasonic flaw detectors can achieve high degrees of resolution by using thousands of samples with sampling frequencies of up to 200 Hz. The output is displayed on the screen in the form of hundreds of minute spots of light known as pixels, each one of which contains a group of perhaps 15 samples. The screen display is changed at a rate of 50 kHz or more so as to provide steady indications. For tests over a short range of, for example, 10 mm the pixel dimension is of the order of 0.05 mm and it is thus possible to obtain a greater accuracy in measurement than with an analogue display, especially as readings are displayed numerically on the screen and thus do not require judgement by eye (Fig. 6.4).

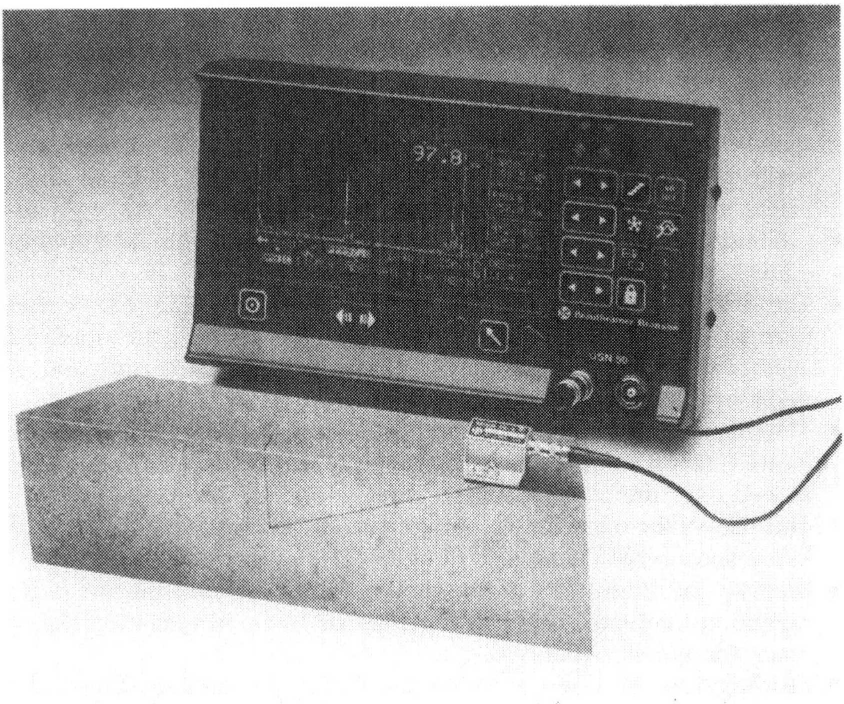

Fig. 6.4 Presentation on the screen of an ultrasonic flaw detector having a high contrast electroluminescent screen display. Reproduced by courtesy of Krautkrämer GmbH & Co.

6.2.5 Advantages and disadvantages of the use of digitized displays

The advantages of converting analogue signals to digital ones are that all of the signals can be stored for future use and that images can be processed either to enhance or suppress certain features. This could be done with analogue signals but with much greater difficulty. Features of digital types of equipment include the following.

- Calibration techniques are built in; the operator is assisted to achieve a correct calibration in one or two easy steps.
- Self-calibration takes place during operation; once the calibration is established, the instrument automatically corrects itself and, if unable to do this, a warning is signalled.
- The measured values are displayed in numerical form, which is of great benefit to the user because he is not required to read the position relative to that of the graticule by eye; this would be

sometimes difficult to achieve accurately. It is also possible to have these readings transferred to a 'head-up' display on a visor, which allows the operator to turn away from the set to concentrate on moving the probe and still see the indications.

- Additional readings on display are surface distance, beam path and depth and the thickness of the plate, when working with an angle probe, provided that the angle is known.
- Calibrations and details of the necessary parameters can be stored in a memory bank and recalled when required.
- The display on the screen can be 'frozen' at any time: the A-scan information remains on the screen until the controls are operated again; this allows the operator to study the screen at leisure and, if necessary, to obtain advice from a supervisor.
- The display can be 'peaked' (the positions of the peaks of the echoes located) so as to build up an envelope of the echoes as the probe is moved over the required range, after which the display is 'frozen'. This allows the operator to comply with the British Standard BS 3923 Part 1 section M4 (1986).
- Because the brightness of the screen is not affected by the pulse repetition frequency, as with analogue displays, it remains constant when the range is reduced.
- Readings can be taken at either the flank, the leading edge of the echo, or the peak. With analogue displays, the standard reading position is the leading edge which, however, moves as the echo height rises and falls due to changes in signal amplitude. On the other hand, with digital displays, the peak height remains constant irrespective of any changes in the value of the signal amplitude and there is thus no need to maintain the height at fixed level (e.g. 80% full screen height).
- The use of noise reduction can ensure that only genuine echoes appear on the screen.
- With the time-corrected gain (TCG) (section 6.2.3), up to 20 reference echoes can be amplified to indicate constant heights over the required range.
- The instrument can be calibrated and then locked to prevent tampering.
- Digitized screen displays can be transferred to video recorders, personal computers and television monitors.
- The use of digitization enables printouts of test data in different languages.
- Interfacing to personal computers and modems, so as to record or analyse results, can be made.
- Circuit boards and extra chips can be easily added to accommodate future accessories.

Disadvantages of the use of digital equipment are few. One is the present-day lack of familiarity with this type of equipment, although this could be overcome by means of a short training programme. Another is the longer response time compared with that of analogue equipment which may be a drawback with the use of automated systems. It should be realized that industry has invested heavily in the more conventional analogue flaw detectors and they are unlikely to be discarded in favour of digital instruments until it is economically feasible to do so.

6.3 ULTRASONIC THICKNESS GAUGES

6.3.1 General considerations

Thicknesses of components in large-size installations such as the hulls of ships and pipe work subject to corrosion in inaccessible surfaces were originally measured by comparatively crude methods such as drilling a hole, using a mechanical depth gauge and then welding a plug into the hole. This method was superseded by the somewhat bulky and heavy pulse-operated ultrasonic flaw detectors which had the further disadvantage of requiring the use of an electric mains supply. However, by the late 1960s they were beginning to be replaced by the lighter and battery-operated continuous-wave frequency resonance gauges as described in section 5.7.2. With suitable calibration, the measured thickness could be indicated by a meter. Further reduction in weight has led to the replacement of the resonance gauge by the lighter, more accurate and sophisticated pulsed instruments, described in section 6.2, providing a liquid crystal display having either a digital or a combined A-scan and digital display.

Operating an ultrasonic thickness gauge is a comparatively simple task for an inexperienced operator but the operator should be properly supervised because of the temptation to feel over-competent in this work. However, for critical surveys, a certificate of competence in ultrasonic testing would be necessary.

Furthermore, the instrument records the time of travel of the waves in the material but the speed of the latter should be accurately known if reliable readings are to be inspected. Thickness gauges are often supplied with a calibration block made from steel in view of the fact that many of their users are testing steel objects. When testing objects made from other materials, allowance should be made for a different value of the ultrasonic velocity; furthermore, the nature of the material may be such that more satisfactory results would be obtained with the use of a different frequency probe depending on the value of the attenuation coefficient (section 3.3).

6.3.2 Testing criteria

A number of factors should be considered when conducting thickness tests.

- The condition of the surface: whether it is flat, curved, rough, scaly or pitted.
- The attenuation properties of the material, which depend mainly on grain size and grain orientation.
- The nature of the reflecting surface of the back wall of the object, which may be eroded and smooth or rough and pitted.
- The temperature of the specimen, which will affect the speed of sound in it. In general, specimens can be tested at temperatures of up to 600°C.
- The nature of the couplant depending on the viscosity; a specially made couplant should be used at higher temperatures.
- The working limits of the probe system, either single or twin, especially when dealing with thin sections.
- The speed of scanning which is limited by the pulse repetition frequency and also the ability of operators to make observations with higher speeds.

6.3.3 Testing procedure

The operation of the ultrasonic thickness gauge involves the measurement of the time taken between the pulse transmission and the receipt of the leading edge, or flank, of the first back-wall echo. To eliminate effects of spurious small echoes from any discontinuities, gating is provided so that echoes having heights of less than a pre-set value are not recorded (Fig. 6.5). To ensure that the amplitude of the back-wall echo does not diminish with distance, a swept gain (time-corrected) control (section 6.2.3) is used. In this way, distance of the leading edge of the pulse from the source does not change in position relative to that of the peak, thus ensuring that the reading of the instrument is the same for all values of gain.

If a single crystal probe is used, it should be mounted on a delay block or shoe which is placed in contact with the near surface of the test specimen; the delay block should be long enough to contain the whole of the dead zone. When a delay line is used, the measured time delay is that between the first echoes from the near and far surfaces of the object being measured. This method is capable of providing high degrees of resolution, ranging from 0.001 mm, for thin 0.15 mm samples, of steel. This technique is limited to testing materials having low attenuation coefficients and parallel surfaces.

The use of single crystal probes is not recommended for measuring the thickness of an object in which the lower surface has been subjected

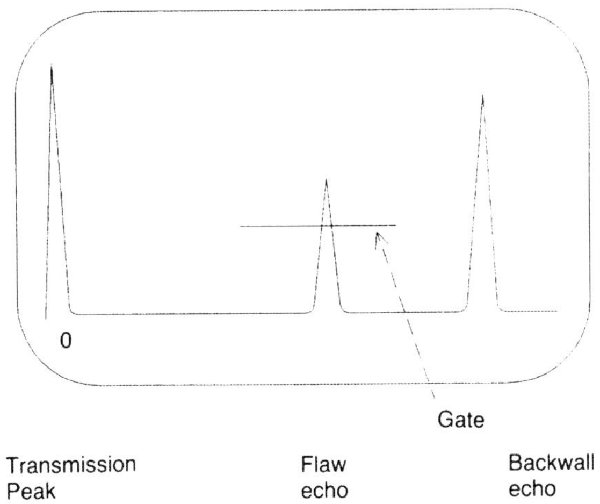

Fig. 6.5 Presentation on the screen of an ultrasonic flaw detector showing the effect of gating to eliminate echoes having heights less than a preset value.

to corrosion pitting. Here, a properly designed twin-crystal probe is used (Fig. 6.6). It has delay lines in which the top surfaces are shaped to produce slightly inclined beams effectively to give rise to 'focusing'. Probes of this type are available with different depths of 'focus', depending on requirements, and careful thought should be given to their selection, e.g. 'near-focused' probes are unsuitable for use with greater depths. This 'focusing' has the effect of magnifying the energy returning from a poor reflecting surface such as one affected by corrosion pitting. The probes, of course, can also detect the presence of this type of defect.

The twin-crystal probe has the great advantage in that problems arising with the presence of a dead zone do not arise. However, it is seen from Fig. 6.6 that, as a result of the inclined beam, the distance

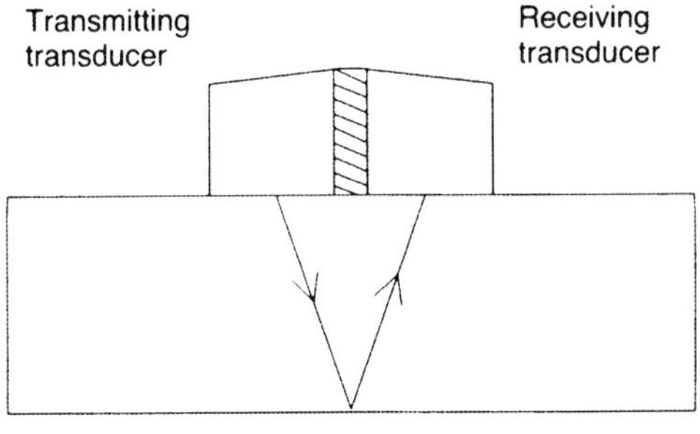

Fig. 6.6 Twin-crystal probe with buffers shaped to create an inclined beam to provide a focusing effect.

travelled by the ultrasound from a twin-crystal probe to produce a single echo is greater than twice the thickness of the test object. It is seen from Fig. 6.7 that, because the beam follows a V-shaped path, the second and higher order echoes are not received and this inhibits calibration. However, this difficulty can be overcome by using a calibration block consisting of steps and placing the probe on each step in turn to obtain thickness variation (Fig. 6.8). The discrepancy between the V-path length s from the probe index to the point of reflection increases with decrease in thickness, t, of the test sample, as shown by:

$$s = [t^2 + (d/2)^2]^{1/2}, \tag{6.1}$$

where d is the distance of separation of the indices of the two probe surfaces. Table 6.1 indicates values of the V-path correction, for d = 4 mm.

The actual errors of measurement for an instrument which has been correctly calibrated are smaller than those given by this table because, when calibrating with a stepped calibration block, the 8 mm step is adjusted to read 8 and the 4 mm step to read 4 and the step thicknesses are thus displayed correctly although the actual beam path lengths are 8.2 mm and 4.5 mm. Provided that the calibration is made within the working range of the measurements, any errors become negligible. Modern microprocessor controlled thickness gauges incorporate automatic V-path correction as well as automatic gain control so as to maintain the returning signal at a suitable amplitude for measurement. The reader should be reminded that these instruments are designed for measurements at fixed points and to operate at a low pulse repetition frequency, typically 3–5 Hz. This precludes the use of scanning techniques, although recently an instrument has been introduced to provide a limited scanning facility so that minimum thickness during the scan is recorded.

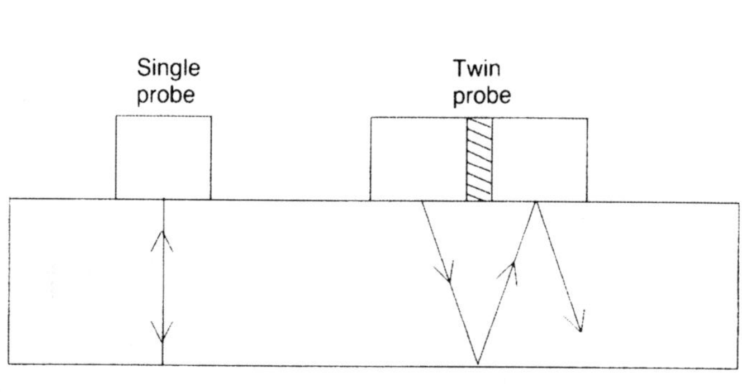

Fig. 6.7 Comparison of ultrasonic paths from a single probe and a twin probe.

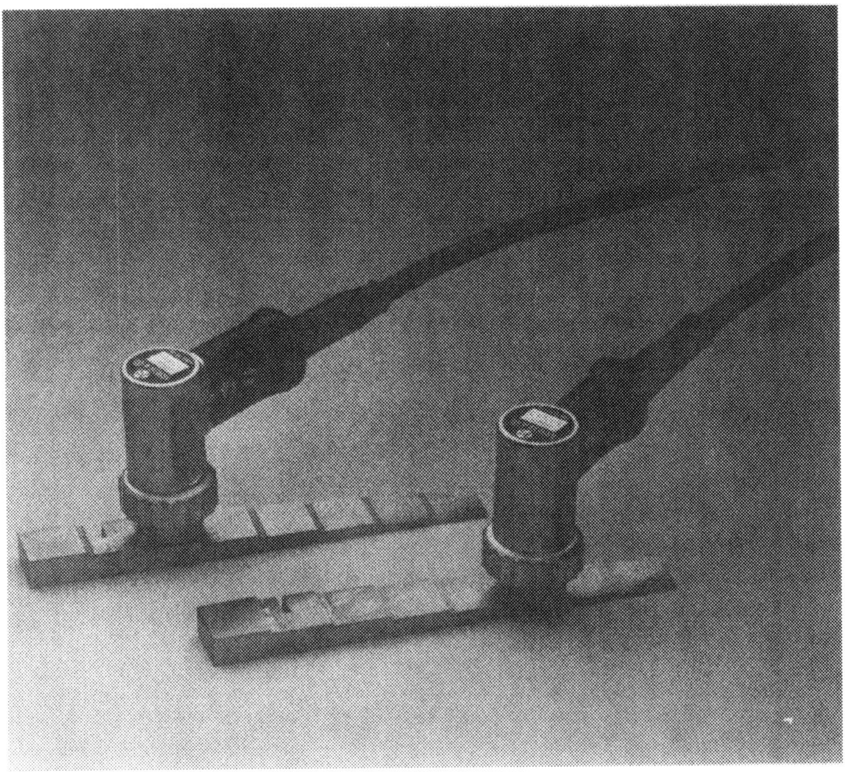

Fig. 6.8 Calibration using step wedges. Reproduced by courtesy of Krautkrämer GmbH & Co.

Table 6.1 Errors in measurement of wall thickness calculated from equation 6.1

Wall thickness t *(mm)*	*Distance along beam,* s[a] *(mm)*	*Measurement error,* e[b] *(mm)*	*Percentage error,* p *(%)*
9.0	9.2	0.22	2
8.0	8.2	0.25	3
7.0	7.3	0.28	4
6.0	6.3	0.32	5
5.0	5.4	0.39	8
4.0	4.5	0.47	12
3.0	3.6	0.61	20
2.0	2.8	0.83	41
1.0	2.2	1.24	124

[a] *s* is the instrument reading for the distance along the beam, which is given the nearest 0.1 mm.
[b] *e* is the calculated value for the measurement error.

6.3.4 Thickness surveys

Thickness surveys, carried out on a regular basis and known as 'on-stream inspection', are conducted on a large scale in chemical plants and refineries whilst they are in operation (Fig. 6.9). The latter is necessary because any shut down involves a great amount of expense, so special thickness gauges have been designed for this purpose (Fig. 6.10). Because the readings must be related to corresponding locations, the gauges are fitted with data-output ports to enable the readings and location information to be passed to a portable high capacity data storage device. The information is passed to a microprocessor which enables it to be stored and displayed or printed out in either graphical or table format.

Thickness surveys are also conducted with bench-model thickness meters when high accuracy is required; instruments of this kind are available with specially designed single-crystal probes fitted with delay blocks to achieve a resolution of 0.001 mm and are capable of measuring thicknesses of steel equal to 0.13 mm. These high degrees of accuracy, which of the order of those achieved by resonance thickness gauges (section 5.7.2) can be influenced by changes in ambient temperature,

Fig. 6.9 Ultrasonic thickness gauge with combined A-scan and digital read-out used for plant survey. Reproduced by courtesy of Krautkrämer GmbH & Co.

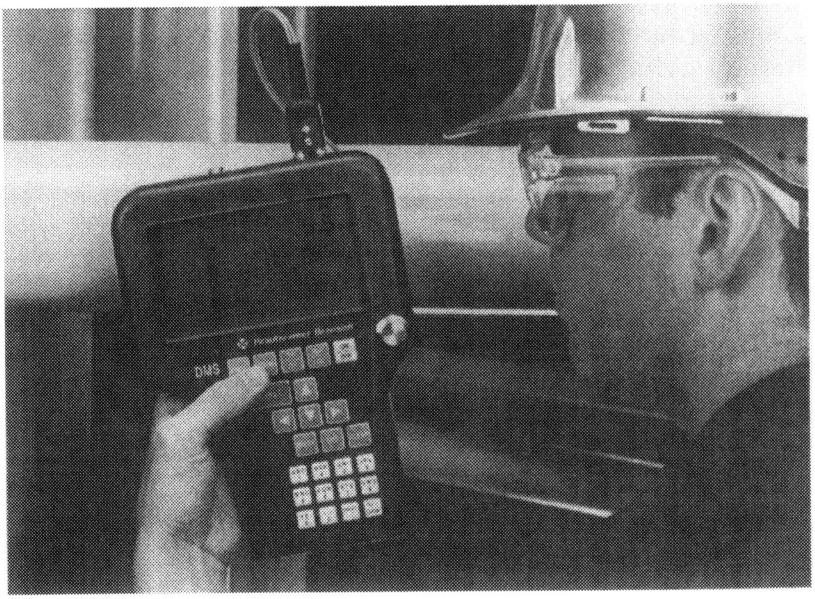

Fig. 6.10 Presentation on the screen of an ultrasonic thickness gauge having a high-contrast electroluminescent screen with a numerical display of thickness reading. Reproduced by courtesy of Krautkrämer GmbH & Co.

which should be suitably controlled. It is also important that calibration blocks are machined to a degree of accuracy consistent with the measurements. These instruments have facilities for switching between imperial and metric measurements and between thickness and ultrasonic velocity values. Further information on testing at elevated temperatures is given by McLaughlin (1989).

6.4 MISCELLANEOUS APPLICATIONS

6.4.1 High temperature testing

When working at high temperatures, typically up to 550°C, the use of probes with heat barrier tips and of high temperature couplants are necessary, even with a contact time of only a few seconds. Too long a contact will cause the probe to overheat and the transducer to depolarize. Furthermore, the increases in temperature in the delay blocks would give rise to changes in indications as a result of the increased ultrasonic velocities. Even if a high temperature couplant is used, it would vaporize in a few seconds and the screen display would consequently vanish. Hence, much practice on the part of the operator is required to acquire sufficient skill to be able to obtain successful readings at the first attempt. Prior to the next reading, it would be necessary to clean the probe before applying a fresh supply of couplant.

6.4.2 Testing painted surfaces

Problems may arise when testing for thicknesses of pipes covered with paint, especially when the thickness of the paint is greater than a few micrometres. A thick coat of paint can act in the same way as a delay block to the probe and the ultrasonic speed in it would certainly be different from that in the material of the pipe. Furthermore, when thick coats of paint are applied, they would generally have varying thicknesses. The use of an interface trigger (section 6.2.3) can help to overcome the latter problem provided that the paint is thick enough to mask the after-ringing of the probe.

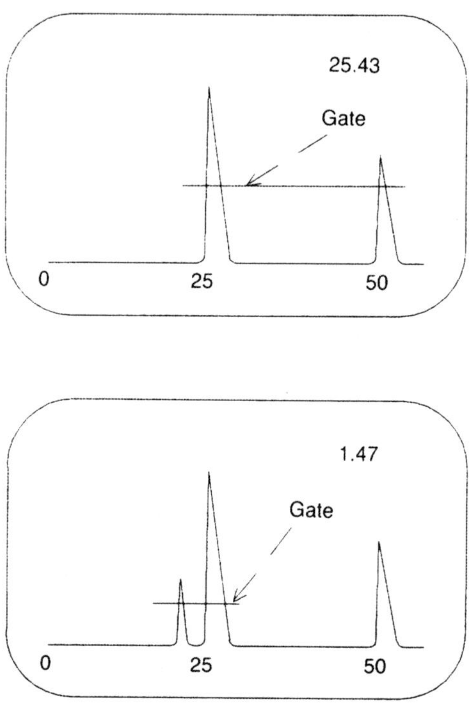

Fig. 6.11 Presentation on the screen of an ultrasonic thickness gauge showing (a) a gate between the first and second back-wall echoes to indicate a thickness of 25.43 mm and (b) a gate between an indication of pitting in the form of a small echo and the first back-wall echo showing a distance of separation of 1.47 mm.

Measurement of the metal thickness is carried out by means of a gate between the first and second back-wall echoes but if the pipe is badly corroded, then the multiple echoes may not be observed and, with a only simple digital display, the operator would not be able to account for this. By using a thickness gauge having both an A-scan display and a digital read-out, the problem could well be resolved. If, for example, pitting were present and an echo in front of the first back echo is observed, the gate can be repositioned in such a way to enable the measurement of the depth of the pit in the back wall (Fig. 6.11). The resultant A-scan display then enables the operator to decide on the most suitable way of measuring the depth of corrosion. Further information on the testing of painted surfaces is provided by McLaughlin (1992).

6.4.3 Advanced ultrasonic flaw detection

When a high level of performance is required for ultrasonic testing, it may be necessary to utilize instruments designed either for laboratory work or for automatic inspection. These instruments are considerably more expensive than those used for everyday testing. The facilities available from them (Fig. 6.12) include multi-channel monitoring in

Fig. 6.12 Automated ultrasonic work station with interlinked computers. (Reproduced by courtesy of Wells–Krautkramer Ltd.)

any desired mode (linear and proportional output), an on-board computer, an interface trigger, a digital or analogue display, an electrically generated distance amplitude correction (section 8.3.2) or time-corrected gain (TCG) (section 6.2.3), trigonometrical functions, data logging, stability and repeatability. In accordance with working conditions, the equipment is either free standing or mounted in racks.

REFERENCES

BS 3973 Pt. 1 (1986) *Manual Examination of Fusion Welds in Ferritic Steels*. British Standards Institute, London.

Highmore, P.J. and Short, K.A. (1988) *Br J NDT*, **30**, 35–9.

McLaughlin, K. (1989) *Br J NDT*, **31**, 196–8.

McLaughlin, K. (1992) *Br J NDT*, **34**, 441–3.

CHAPTER 7

Ultrasonic flaw detection

7.1 INTRODUCTION

A wide variety of methods for ultrasonic testing is discussed in Chapter 5 but the pulse-echo technique (section 5.3) is, by far, the most commonly used method for flaw detection. Flaw detection is usually applied to testing metals, so it is not surprising that most of the relevent national and international standards are concerned with testing metals using the pulse-echo technique. These standards are principally concerned with the design and calibration of the equipment used and the choice and applications of probes, which are discussed in this Chapter and in Chapter 9. Also included here are considerations of problems encountered with forgings, particularly related to the difficulties brought about by mode conversion between transverse, surface and longitudinal waves and the use of probes for immersion testing.

7.2 CALIBRATION OF EQUIPMENT

7.2.1 General considerations

Calibration of an ultrasonic flaw detector should initially be carried out either by the manufacturer or by a service engineer and guidance is given, for example, by British Standard BS 4331 Part 2 (1982). This calibration is essentially electrical and requires the use of specialized equipment.

Calibration should also be made by the operator either daily, weekly or monthly as detailed in British Standards BS 4331 Part 1 (1978, 1989) and Part 3 (1974, 1987), making use of the blocks described in section 7.2.2. In the USA, ASTM E317-85 (1985) is used for the calibration of pulse-echo testing equipment. The calibration checks, which are dealt with here, are essential for the correct usage of the equipment and to detect any faults which would require the use of a service engineer. They comprise the following.

- Linearity of time base (sweep length): weekly.
- Time base calibration: daily or at a change in the nature of a test.
- Linearity of amplifier gain: weekly.
- Probe index for angle probe: daily.
- Beam angle for angle probe: daily.
- Sensitivity and signal-to-noise ratio: daily per probe.
- Beam alignment (squint) for angle probe: daily.
- Beam profile: monthly.
- Dominant frequency: monthly.
- Pulse length: daily per probe.
- Dead zone: monthly.
- Near field: monthly.
- Signal-to-noise ratio: monthly.
- Overall system gain: daily or after repairs.
- Resolving power (depth): monthly or after repairs.
- Resolving power (angular): monthly.

The first seven of these checks are performed on site (BS 4331 Part 1) and the remaining checks for the monitoring of probes, when in service (BS 4331 Part 3).

7.2.2 Calibration blocks

The blocks suitable for calibration of ultrasonic equipment are described by British Standard BS 2704 (1978) and their specifications are given by the international standards ISO 2400 (1972) and 7963 (1992). They are usually made from either low- or medium-carbon ferritic steel which has been deoxidized by means of additives such as manganese or silicon before casting, and normalized after rough machining with the use of heat treatment at a temperature of 800°C or higher. This refines the grain size and produces a uniform structure. Final preparation is provided by light grinding with an average surface roughness value R_a equal to 0.8 μm. British Standard BS 1134 (1988) describes how the surface roughness is measured.

Three calibration blocks, referred to here by the British Standards nomenclature A2, A4 and A5 (BS 2704 1978, 1983) are in common use. The A2 block (Fig. 7.1) is known also as the International Institute of Welding or IIW block, for which the international standard is ISO 7963 (1985). The European CEN standard, which is identical to ISO 7963, is known as EN 279633 (1992) and the German standard as DIN 54120 (1966, 1973). The A4 block (Fig. 7.2) is sometimes called the V2 block and the A5 block is often referred to as the Institute of Welding (IOW) block. A wide variety of test blocks, exceeding fifteen in number, is available in the USA.

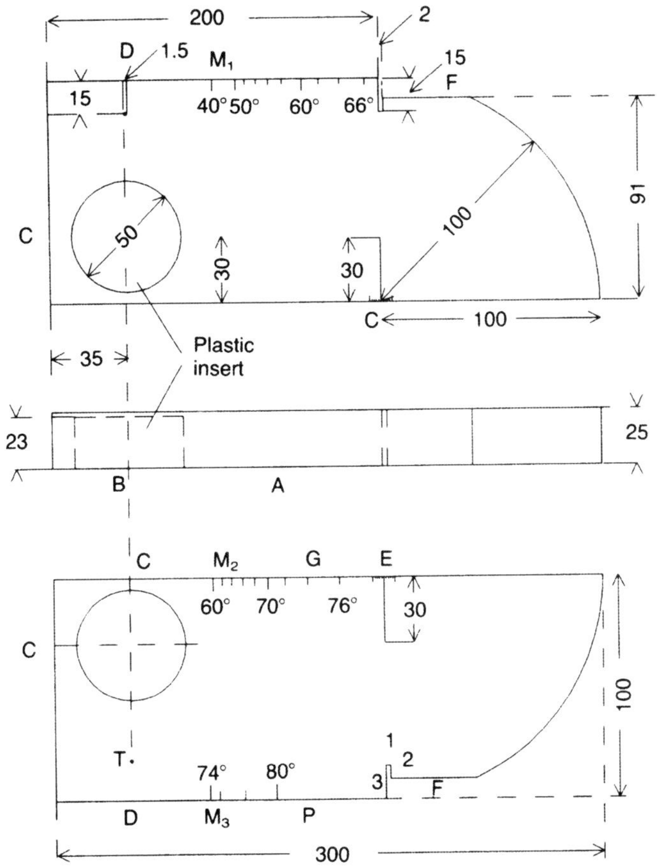

Fig. 7.1 A2 (or IIW) calibration block. Numerals indicate distances in millimetres and letters show the positions for compression wave probes for standard calibrations: A for multiple echo calibration in the thickness direction (50 mm), B for echoes from the plastic insert having a thickness equivalent to that of a 50 mm thickness of steel, C for dead-zone testing, at either of the 10 or 5 mm distances, E for resolution testing in which echoes are received at distances of 85, 91 and 100 mm, F for the use of shear wave calibration with a compression wave probe, for which the respective ultrasonic speeds bear a ratio of 50 to 91 and G for multiple echo calibration along the 100 mm width direction which indicates side-wall echoes due to beam spread. (After British Standard BS 2704, 1978, 1983)

7.2.3 Calibration procedures

Linearity of time base

The linearity of the time base, sometimes called the sweep length, can be checked by means of a compression-wave probe located at the position A on the largest plane surface of the calibration block (Fig.

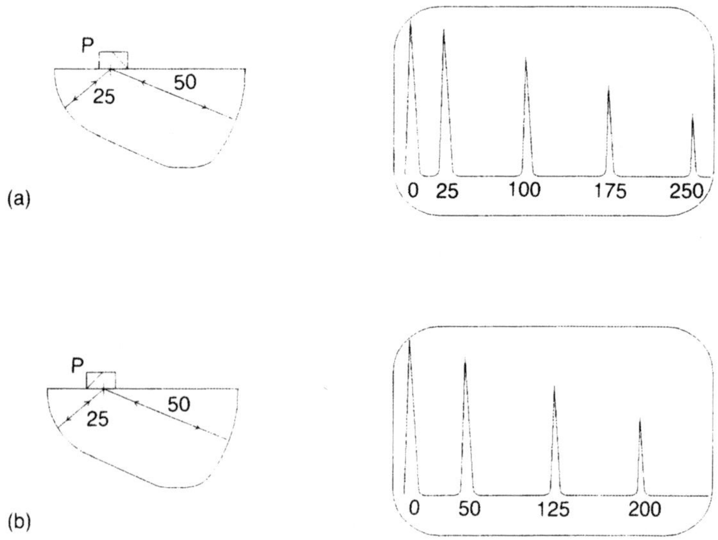

Fig. 7.2 A4 (or V2) miniature calibration block with angle probe directed along (a) the 25 mm radius and (b) the 50 mm radius with corresponding echo signals. Numerals indicate distances in millimetres. (After British Standard BS 2704, 1978, 1983).

7.1), for which the thickness is sufficiently small, i.e. 25 mm, to allow for, say, 10 multiple echoes to appear on the instrument's screen when a suitable range is chosen. The time base is then adjusted so that the leading edges or flanks of the first and 10th backwall echoes coincide with the appropriate scale markings on the graticule. The height of each echo is adjusted to approximately 80% of the maximum screen value and the deviations in the positions of their leading edges from those of their corresponding scale markings are noted. This height is essential to ensure that the relative positions of the leading edges are the same for each of the observed peaks. The time base is regarded as being linear if all of the deviations are less than ±2% of its total range.

With some more recently manufactured equipment (section 6.2.5) the position of the echo is indicated by the peak rather than the leading edge, or flank. It is no longer necessary to adjust the echo heights, because the positions of the peaks remain the same for all values of height.

Calibration of the time base

With time-base calibration for compression waves, the probe is placed on one of the parallel surfaces of the calibration block in a position at which only echoes from the opposite parallel surface are observed. The block should be made from a material in which the ultrasonic speed is known, e.g. the IIW test block, or made from the same material as that of the object being tested. Equally spaced multiple echoes are observed on the oscilloscope screen on excitation of the probe. For example, with the probe at position A (Fig. 7.1), if the chosen range is 100 mm, the calibration block is 25 mm thick and its material is the same as that of the test object, four multiple echoes each indicating a 25 mm thickness are observed. Here the reader is reminded not to confuse thickness with ultrasonic path length, which has twice the value. With a single-crystal probe, the initial pulse is displayed at the zero position. Using the fine range and delay controls (section 6.2.3), the four back-wall echoes are adjusted in position until they coincide with the 2.5, 5.0, 7.5 and 10.0 indications on the graticule on the screen of the instrument (Fig. 7.3). It should be noted that the distance between the first back-wall echo and the first transmission signal is somewhat longer than each of the equal distances between the subsequent reflected echoes because of the physical characteristics of the probe, e.g. the effect of the protective cover of the transducer, so that its position is not indicated by the zero indication on the graticule.

This also arises with probes having delay blocks and with twin-crystal probes (section 4.4). When these four echoes are properly

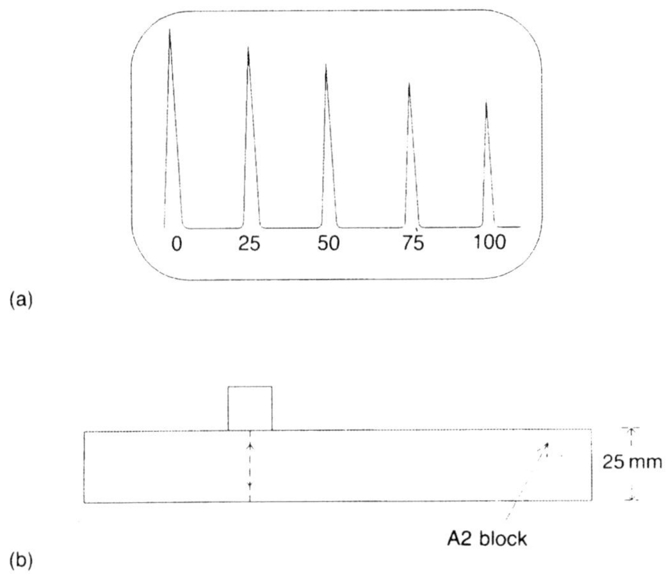

Fig. 7.3 Locations of (a) backwall echoes with (b) probe located on the surface of an A2 calibration block. Numerals indicate divisions on the time base of the screen of the instrument.

located, the initial pulse will have drifted into its correct position, known as the true zero point, and the calibration is complete. The twin-crystal probe method may be similar but, because the delay blocks are in front of the crystal, the position of the initial pulse lies before the zero point and would thus not be visible at this point. Great care must thus be taken to locate multiple and separate echoes correctly. The absence of an initial pulse provides an advantage for dealing with near-surface echoes and this is the main reason for the use of twin-crystal probes.

For shear or transverse wave calibration, an angle probe is located at the centre of an arc of a circle having a specified radius. With the IIW (A2) block (Fig. 7.1) the centre is at point H and the radius is 100 mm but, with this arrangement, it is not possible to obtain multiple echoes for ranges of 200 mm or greater. Using the delay control to reset the first backwall echo to zero and the second one to 100 mm, the 100 mm range can be calibrated. With the A4 block (Fig. 7.2), arcs having radii of 25 mm and 50 mm are provided and, with the probe located at point A, echoes can be obtained at distances of either 25 mm or 50 mm from this point depending on which direction the probe is facing. It is easy to see that, for one direction of the probe, the first echo corresponds to a distance of 25 mm and multiple echoes are observed corresponding to distances of 100 mm, 175 mm, 250 mm etc. and, for the other direction, the first echo corresponds to a distance of 50 mm and the multiple echoes to distances of 125 mm, 200 mm, etc.

It is possible to obtain regularly-spaced echoes corresponding to shear waves if one is prepared to assume that the time taken for compression waves to travel a distance of 91 mm is equal to the time taken for shear waves to travel 50 mm, which is the case for the type of steel used for the IIW test block. The calibration is performed by locating a compression wave probe at the point F (Fig. 7.1) and thus avoiding the need to use 'radiused' blocks.

It should be pointed out here that some of the latest **digitized flaw detectors** possess either semi- or fully-automatic systems for time base calibration, together with a memory-storage facility.

Linearity of amplifier gain

With a linear amplifier, the height of an echo signal on the oscilloscope screen should vary with the amplitude of the corresponding received echo in a linear manner: a doubling of the amplitude should result in a doubling also of the signal height. With a logarithmic amplifier, the height of the observed echo signal should increase by a constant amount whenever the amplitude is doubled. Testing for linearity involves the testing of the amplifier and the gain control, which is calibrated in decibels. It is important to ensure that all of the controls of the flaw detector are in the positions required for testing and with the suppression/reject and swept-gain controls switched off.

The methood of testing involves positioning the probe on a calibration block and maximizing a suitable echo signal to 80 % full screen height (FSH). The reading in decibels of the gain control is noted. The gain setting is then increased by 2 dB for which the signal amplitude should have risen to 100 % FSH. Any deviation from this value should be noted and the gain setting should then be altered to its original value of 80 % FSH. The gain should then be reduced by 6 dB for which the signal height should be 40 % FSH and all deviations again noted. On reducing the gain by a further 12 dB, the signal height should be 10 % FSH and any deviation noted. On a reduction of another 6 dB, the signal amplitude should be 5 % FHS but it is acceptable for the signal to be visible above the base line. Each of these deviations should be taken into account for all subsequent tests.

Probe index for angle probe

The probe index is the point at which the centre of the beam leaves from and returns to the probe shoe (section 4.4.1) and is the point determining the position of the probe. However, it may shift in position as a result of wear of the lower surface of the shoe. Both the blocks A2 and A4 are suitable for locating this point. With the A2 block, the probe is located at point E, the centre of the graduated scale, to propagate towards the arc and it is then moved along the scale in either direction to obtain maximum amplitude of the reflected echo. If the probe index has moved from the zero position of the scale a mark should be made on the shoe to indicate its new position.

Beam angle for angle probe

The angle characterizing an angle probe is that of refraction in the object being tested (section 2.5.2), which may change when propagation takes place into another material. This angle can be checked using either the A2, A4 or A5 block but it can be done more accurately with the larger A2 block (Fig. 7.1) on which the graduations are at only 2° intervals as opposed to 5° intervals with the A4 block. With the A2 block, the probe shuld be located on one of the scales M_1 (40° to 66°), M_2 (60° to 76°) or M_3 (74° to 80°), depending on the value of its angle. With the probe on either the M_1 or M_2 scale, the beam is aimed at the periphery of the plastic disc but if it is on the M_3 scale, it is aimed at the small hole (Fig. 7.4). The probe position is then adjusted until the maximum signal is observed and its angle determined from the position of the probe index on the scale, corrected if necessary from the last operation.

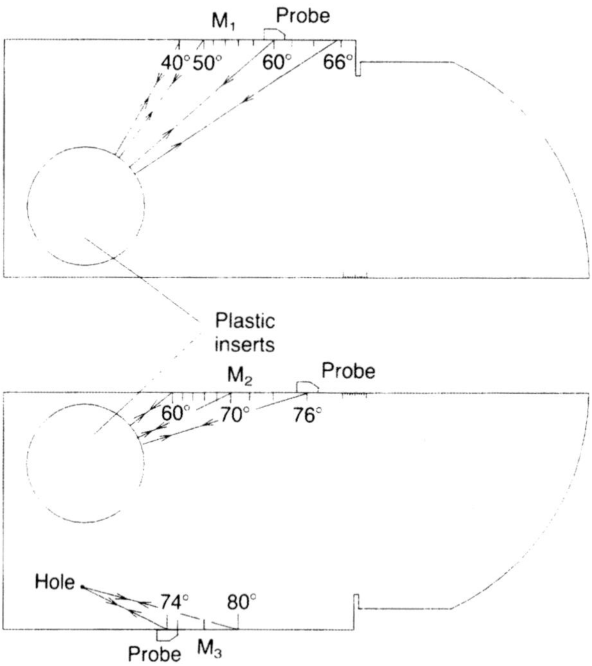

Fig. 7.4 Locations of angle probes for checking beam angles. The probe is placed on one of the scales M_1 and M_2 for reflection at the 50 mm diameter plastic insert or on the scale M_3 for reflection at the 1.5 mm diameter hole. Subdivisions of the scales are at 2° intervals but, for geometrical reasons, they are not equally spaced apart.

When the position of the probe index lies between two graduations, it may, of course, be determined by interpolation but greater accuracy can be achieved using a trigonometrical method. Figure 7.1 indicates that the centre of the 50 mm diameter plastic insert in the A2 block is located at 70 mm from the 200 mm side, 30 mm from the 300 mm side and 35 mm from the 100 mm side. When the probe is positioned on the longer of these sides, the distance, d, in millimetres of the point A (Fig. 7.5) from the 100 mm long side, for maximum echo, is measured with a ruler. Considering a point B located 35 mm from the end, a right-angle triangle ABC, where C is the centre of the insert, is defined for which the angle BCA is clearly seen to be equal to the probe angle α. Hence:

$$\tan \alpha = (d - 35)/30.$$

When it is necessary for the probe to be located on the longer of these sides, we have:

$$\tan \alpha = (d - 35)/70.$$

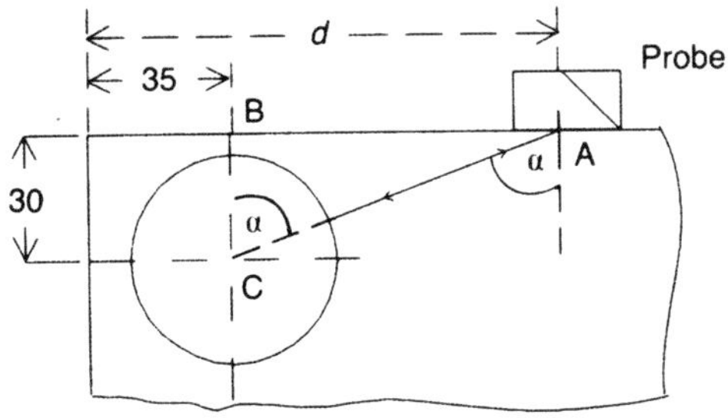

Fig. 7.5 Trigonometrical method of measuring probe angle α. d = distance of the probe index A from the edge of the A2 block.

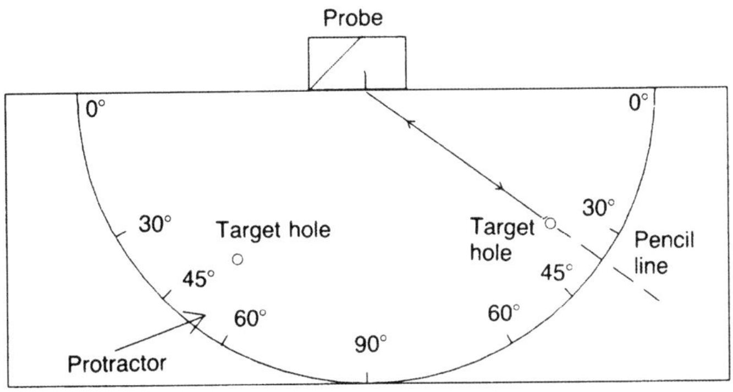

Fig. 7.6 Use of A5 (IOW) block for measuring the probe angle with the aid of a protractor.

With other blocks, such as the A5 (Fig. 7.6), the target takes the form of a small hole, of 1.5 mm diameter, drilled horizontally in a suitable position in the far field. The position of the probe is adjusted until the echo is maximized. With the aid of a ruler, a line is then drawn from the edge of the block at the index point to the hole and the angle between this line and the normal, which is measured with a protractor, is equal to the probe angle.

Sensitivity and signal-to-noise ratio

The main object of this test is to provide a simple method for checking the sensitivity of the probe and the flaw detector. This method is not intended as a method of defining working sensitivity which should be set according to the job requirements and the testing standards.

A standard calibration block (A2, A4 or A5) containing a 1.5 mm hole for a target may be used. The sensitivity is checked with the relevant flaw detector controls set to the positions to be used in subsequent testing. The procedure is as follows.

1. Place the probe on the chosen calibration block and maximize the echo from the hole to be used as a sensitivity check.
2. Adjust the gain control to set the echo to 20% FSH and note the setting in decibels.
3. Increase the gain setting until the overall noise is at the same 20% level at the same distance as the target hole, note the new setting.

The first measurement (2.) provides a check on the sensitivity of the probe and flaw detector. The difference between (3.) and (2.) gives the signal-to-noise ratio.

Beam alignment of shear-wave probe (squint)

Testing for beam alignment, or probe 'squint', is necessary to ensure that the axis of the beam coincides with the line joining the index point with the target. To measure the angle of 'squint', it is necessary to use a smooth steel block with a square edge and the IIW (A2) block is ideal for this purpose. The probe is located in the position shown in Fig. 7.7 to ensure that the beam is reflected from the bottom edge of the block: at half skip distance. The probe is then swivelled until the signal is maximized and a ruler placed alongside the shoe of the probe towards

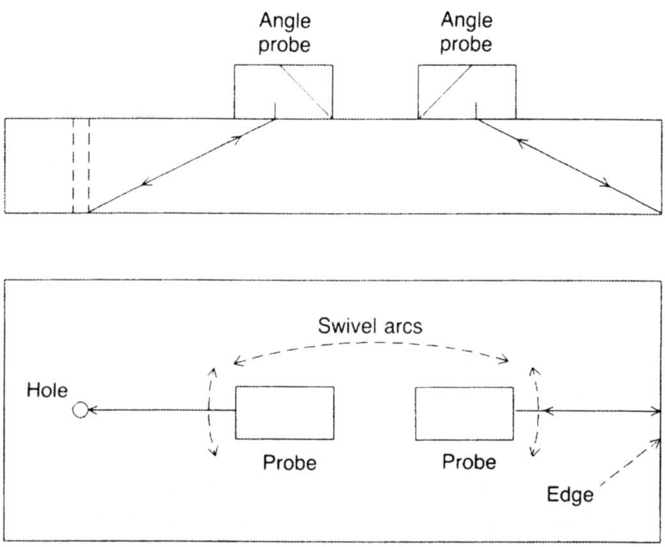

Fig. 7.7 Measurements of the angle of squint of an angle probe using, as targets, (a) a drilled through-hole and (b) a plate square edge.

the edge of the plate and a line drawn to intersect latter. The angle between this line and a line drawn perpendicularly to the edge at the point of intersection is the angular deviation or **squint** of the acoustic axis of the beam.

Beam profile

Special profile blocks are recommended for plotting beam profiles for both compression- and shear-wave probes. Both plane surfaces and cylindrical holes are used as targets in these blocks, depending on the purposes of the plots; for example, plane reflectors are used when assessing laminations in steel plate.

Targets should be located at suitable depths in order that the minimum and maximum working ranges for the intended beam plot are obtained. The shear-wave probes are tested with the side of the IOW block A5, utilizing the 1.5 mm cross-drilled holes. The method of producing this beam plot is described in section 8.2.3.

Dominant frequency

No recommendations have been made for the assessment of the dominant frequency, usually the fundamental frequency, for a rectified display on the screen; for an unrectified display, the trace of a pulse which has been reflected from a small target is brought into the calibrated section of the time base with the leading edge coinciding with an appropriate graticule division. The dominant frequency is then estimated by counting the number of cycles occurring in a specified time interval, e.g. 1 μs.

Pulse length (or width)

The pulse length or pulse width is defined here as the time expressed in microseconds taken to reduce to 10% of its peak value. In the absence of a microsecond display, the time lapse is estimated from the corresponding distance on the calibrated time base. For highest accuracy, a short test range should be used.

Dead zone

The dead zone, which often provides difficulties with single crystal compression-wave probe operations, is brought about by the 'after ringing' of the initial pulse (section 5.3.1). This gives rise to intrusion into the 'working area' of the test object with the results that small flaws are not detected with certainty in this region which lies adjacent to the probe. The extent of this zone is estimated either by observing the calibrated time base or with the use of the IIW (A2) block (Fig. 7.1) or by locating one of the two positions C for which respective target holes of 5 mm and 10 mm target are found.

Near field

The near field or Fresnel zone (section 2.7.1) for a transducer extends from its surface to a distance of $D^2/4\lambda$ (r^2/λ) where D and r are, respectively, the diameter and radius of the transducer and λ the wavelength which is equal to c/f where f is the frequency and c the speed of ultrasound in the material being tested. Figs. 2.11(a) and (b) indicate wide fluctuations of acoustic pressure and, hence, of sensitivity of testing in this region. As shown by these diagrams, sensitivity is at a minimum at the mid-point of the near field, and specially designed test blocks with flat bottom holes are available for assessing the size of the near field (British Standard BS 4331, Part 3 (1974, 1987)). Unless a highly accurate assessment of its length is required, the distance can be calculated from the expressions given above.

Signal-to-noise ratio

The test for signal-to-noise ratio is important, if it is necessary to ensure the detection of small defects. The following procedure is adopted.

1. Switch off the suppression or reject control of the equipment.
2. Apply the probe to the workpiece and set the back-wall echo signal to a selected height, 20 % FSH.
3. Note the level in decibels of the attenuator, adjust it until the noise level reaches the same height as that of the signal and note the new attenuator level.

The difference between these two readings of the attenuator is thus the signal-to-noise ratio.

Overall system gain

A probe is placed on an IIW (A2) test block at point G (Fig. 7.1), so as to target on the 50 mm diameter plastic insert, and the number of echoes, usually two or three, is observed. Any deterioration in the echo pattern would be caused by faults in either the probe or the equipment. The question of which of these is faulty can be resolved by replacing the probe with a standard one kept especially for this test.

Resolving power

The resolving power of an ultrasonic flaw detector is a measure of its ability to distinguish between neighbouring objects. For compression wave testing, the resolution can be tested by locating the probe at point E on the A2 block (Fig. 7.1) and aimed at the three reflecting faces 1, 2 and 3 which vary only slightly in depth. For good resolution, the echoes from each of these faces are clearly identified but, if the resolution of the probe is poor, they cannot be distinguished from one another (Fig. 7.8(a)).

The resolution of an angle probe can be determined with the use of a modified A5 block containing additional holes (Fig. 7.8(b)) which are set at an angle of 10° to the vertical; the top two holes are spaced 4 mm apart and the remainder are 2.5 mm apart. Alternatively, a BS 2704 A7 block specially created for testing resolutione may be used (Fig. 7.9). It is semi-cylindrical in shape and formed, by turning from a 150 mm

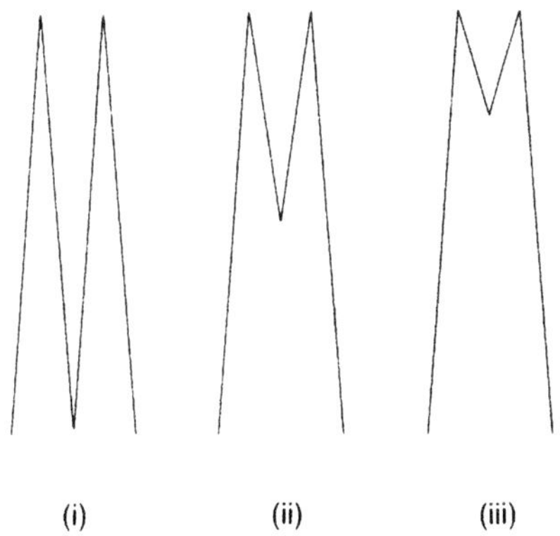

(a)

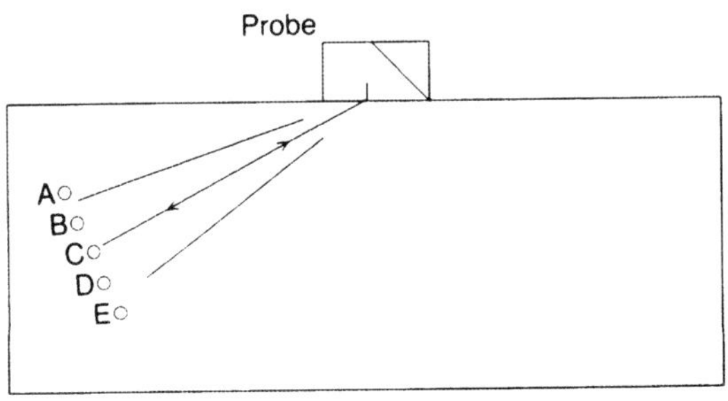

(b)

Fig. 7.8 (a) Screen indications of resolutions of echoes: (i) 100 %; (ii) 50 %; and (iii) 25 %. (b) A5 (IOW) block modified, by drilling holes, to check the resolution of angle probes. Separations of the holes are 4 mm for A to B and B to C and 2.5 mm for C to D and D to E.

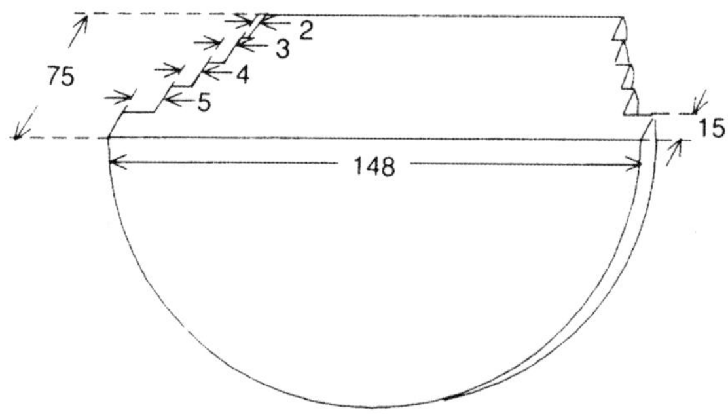

Fig. 7.9 A7 test block, as used for measuring the degree of resolution of an angle probe. Numbers denote distances in millimetres.

diameter bar, with diameters reducing in steps of 1 mm from 5 to 2 mm. The probe is placed, pointing in a radial direction, on the flat surface at the centre of this diameter and moved in an axial direction so as to target the progressively reducing circumferences.

The resolving power of a probe depends on the frequency and on the type of waves propagated. At given frequency, shear waves have about half the wavelength of compression waves and thus the corresponding pulse width is reduced by about one half, so that their use provides a higher degree of resolution than using compression waves. Because the wavelength decreases with increasing frequency, greater resolution can be achieved by increasing the frequency.

Angular resolving power

The test for angular resolving power is the same for that for the beam profile but its purpose is different. Here a probe which provides the smallest angle of beam spread for a given drop in decibels (e.g. 6 or 20 dB) is selected from a batch. The index, θ, of angular resolution is given by $\tan\theta = D/R$, where D is the lateral probe shift providing the appropriate decibel drop and R is the range, i.e. the distance between the probe index and the reflecting surface.

7.3 ANGLE PROBE OPERATION

With the pulse-echo method of ultrasonic testing, it is usually desirable for the sound beam to be incident at right angles to a defect and this would present no problem when the defect lies parallel to the surface of the object being tested, for which normal compression waves are used. Difficulties arise with inclined defects which would require the use of angle probes generally propagating shear waves. However, one should take into account the fact that, because these waves are usually in the far field when they reach a defect, there is a continuous spread of incidence angles and, provided that the angle of the probe is properly chosen, some of the rays in the beam will be incident at virtually right angles to the defect.

Angle-beam testing is common in the fabrication industry for the inspection of welded plates, pipes, tee pieces, cruciform and nozzles, which are all commonly made from steel. The welds vary in type which include single vee, double vee, lap, U prep and J prep (Fig. 7.10) with different angles of preparation of the fusion face, and great difficulties are presented to the tester. As a result, not many operators are capable of testing all of these configurations and many of them specialize in only one type of joint preparation. Consequently, certifications issued to weld testers should specify the categories in which they have gained approval.

The angle probes most commonly used are the 45°, 60° and 70° probes, for which the angle of incidence is greater than the critical angle for Perspex and steel (section 2.5.2), with the result that only shear waves are transmitted. Two other standard probes, 35° and 80°, exist but they are used less frequently. With the former, the spread of the beam in the shoe may give rise to transmission of some compression waves incident at an angle below the critical angle and, with the latter, the beam spread may give rise to the transmission of mode-converted surface waves.

An important quantity connected with the testing of plates with shear waves is the **skip distance**, which is defined as the distance in front of the probe at which the beam emerges after reflection from the lower surface. If t is the thickness of the plate and α the probe angle, it can be seen from Fig. 7.11 that the skip distance, d, is given by:

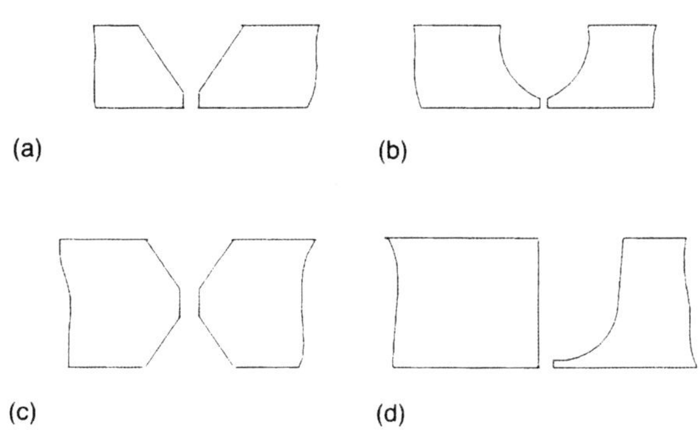

Fig. 7.10 Types of weld preparations, (a) single vee, (b) double vee, (c) U prep and (d) J prep.

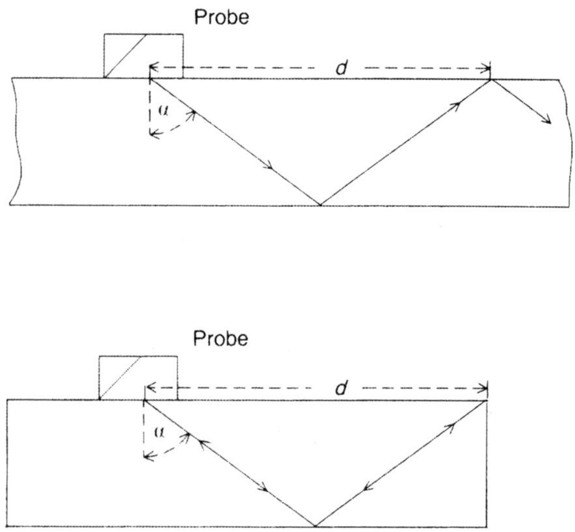

(a)

(b)

Fig. 7.11 Skip distance, d, for shear-wave probe of angle α showing reflections (a) away from the edge and (b) at the edge.

$$d = 2t \qquad \tan\alpha = ft, \tag{7.2}$$

where

$$f = 2\tan\alpha.$$

Here f is the **skip factor** and it has values of 2.0, 3.5 and 5.5 for respective probe angles of 45°, 60° and 70°. Figure 7.11 shows that when a probe is placed at one skip distance from the edge of a plate, the echo occurs at the top corner and when at a half skip distance from that edge, it occurs at the bottom corner. Thus with a weld in a single-wall plate or a pipe in the longitudinal direction on a circumferential seam, the cap can be tested from full skip distance and the root from half skip distance. Figure 7.12 shows that, when a probe is used to test the longitudinal seam in a pipe, the skip distance is longer than that required to test a plate having the same wall-thickness. This factor is discussed below.

Testing tubular products with an angle beam probe in a direction parallel to the axis presents few problems. However, when the tests are made with the probe passing around the external circumference, it is essential that the values of the outside diameter, D, wall thickness, t, and probe angle, α, are suitable to allow the beam to strike the inside surface at grazing incidence. Figure 7.13(a) illustrates that when the

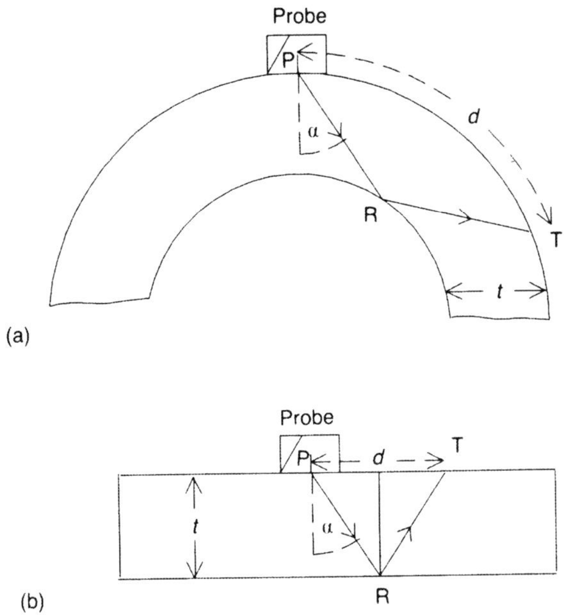

Fig. 7.12 Comparisons of skip distance *d* and path length PRT for propagation of shear waves in (a) a pipe and (b) a plate made from the same material and of equal thickness.

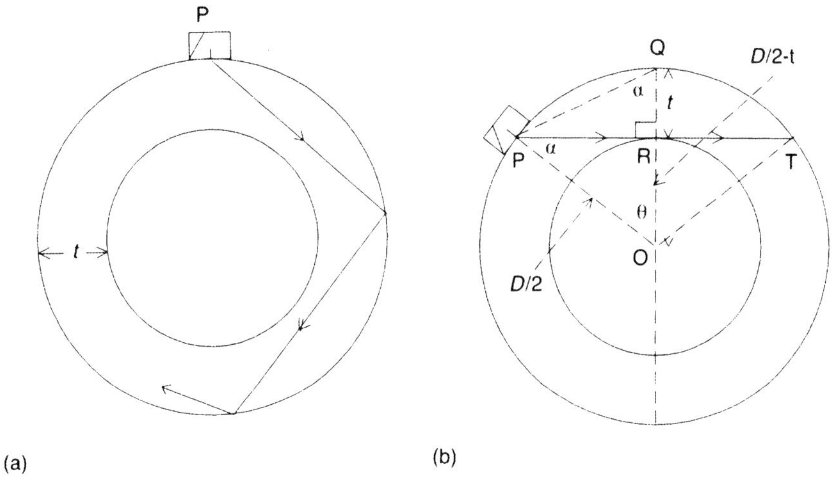

Fig. 7.13 Propagation of shear waves by a probe P into a pipe of wall-thickness, *t*, and diameter, *D*, when the ratio *t*/*D* (a) is too high to allow reflection at the inner surface and (b) grazes the inner surface. The latter figure has been rotated anticlockwise by an angle θ to clarity the symmetrical arrangement.

probe angle is greater than α, reflections occur only at the outer surface, i.e. the waves do not reach the inner surface of the pipe. Figure 7.13(b) illustrates what happens when the probe angle is equal to α, which is the maximum angle for which the beam strikes the inner surface at R, at 90° incidence. It is clearly seen that, for the triangle PRO, the value of $\sin\alpha$ is equal to OR/OP = $(D/2 - t)/(D/2)$, from which it follows that:

$$t = D(1 - \sin\alpha)/2. \tag{7.3}$$

$F = t/D$, in this context, is called the **probe factor** and Table 7.1 provides values for different values of α. Thus, for example, if the calculated value of F is 0.140, the probe angle should not exceed 45° if the beam is to be reflected from the inner surface of the tube. In practice, the lowest limit for probe angle is about 35° which is set by the value of the critical angle for transmission of compression waves from the probe shoe, usually made from Perspex, into the steel sample (section 2.5.2).

A characteristic of this type of angle beam propagation in pipes is that the skip distance and beam path length are greater than those for a flat plate having a similar thickness, t (Fig. 7.12(b)). Figures 7.14 and 7.15 illustrate curves (Krautkrämer and Krautkrämer, 1990) relating the path length and skip distance for a pipe wall to those for a plate having the same thickness. From these curves, one can derive factors f_s and f_p which indicate the ratios of skip distance to sound path in the wall of the tube and plate, respectively, for different values of d = t/D. It is shown in Appendix D that:

$$f_s = \theta\pi/[360d\tan\alpha], \tag{7.4}$$

and

$$f_p = (1 - 2d)\sin\theta/(2d\tan\alpha), \tag{7.5}$$

where

$$\theta = \sin^{-1}[(\sin\alpha)/(1 - 2d)] - \alpha. \tag{7.6}$$

These equations are, of course, subject to the conditions imposed by equation 7.3.

The curves shown in Figs. 7.14 and 7.15, correspond to probe angles having values of 35°, 45°, 60°, 70° and 80°. Curves for probes having other angles can be interpolated between appropriate existing ones. The curves end abruptly when the value of $\sin\alpha$ at the points beyond which values are no longer obtainable because the probe angle is too large to allow the sound path to reach the inner surface of the pipe and

Table 7.1 Values of probe factor, F, for different values of probe angle as calulated from equation 7.3

Probe angle, α (degrees)	35	45	60	70	80
Probe factor, F	0.213	0.146	0.067	0.030	0.0076

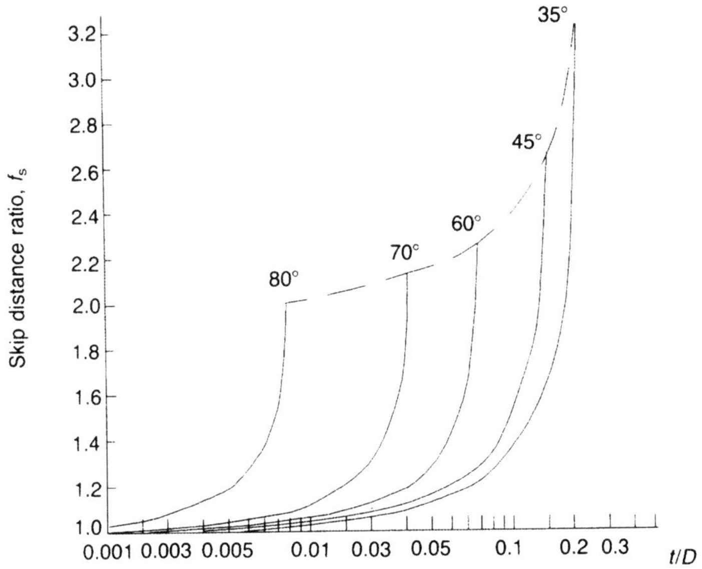

Fig. 7.14 Relationship between the ratio of skip distances f_s and t/D for pipe and plate, respectively, when t is the common thickness and D the outer diameter of the pipe (after Krautkrämer and Krautkrämer 1990).

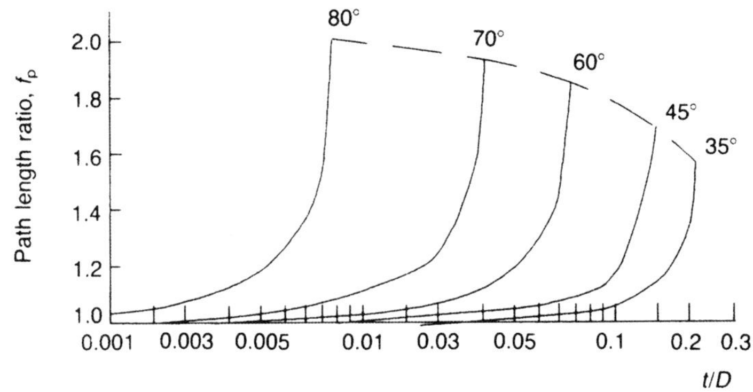

Fig. 7.15 Relationship between the ratio of path lengths f_p and t/D for pipe and plate, respectively, when t is the common thickness and D the outer diameter of the pipe (after Krautkrämer and Krautkrämer 1990).

reflection does not take place. These points are joined by the curve represented by broken lines.

As an example of the application of these curves, consider the use of a 45° probe to test a pipe having a wall thickness of 45 mm. The respective values of the skip distance and beam path for a plate of this thickness are 90 mm and 127 mm. If the pipe has an external diameter of 300 mm the value of t/D is 0.15. From the graphs, the values of f_s = 2.6 and f_p = 1.65 are obtained so that, in the pipe, the skip distance is 90 × 2.6 = 234 mm and the sound path is 127 × 1.65 = 210 mm. These values are considerably longer than those for the flat plate, but their differences have been reduced from roughly 30% to 10%.

7.4 THE TESTING OF FORGINGS AND OF ROLLED AND WROUGHT METALS

Forgings and rolled and wrought products are obtained from a cast metal ingot (Fig. 7.16) and may still contain many faults even after the invariably defective top section has been discarded. Faults left in the ingot may include primary and secondary pipe, A and V segregations and porosity. An account of these has been provided by Taylor (1988).

When an ingot is reheated and then forged or rolled, the original faults are flattened and drawn out from their original shapes. For example, any porosity changes from roundness to flatness and the resulting defect may then be classed as an inclusion and any primary or secondary pipe would take the form of a lamination; rolled plates can thus contain flaws which can be classified as being either 'large' or 'small' and these terms can be related to the size of the ultrasonic beam. 'Large' defects are regarded as those having dimensions greater than the diameter of the testing probe, usually 20–25 mm. Inclusions in the ingot are regarded as being 'small' defects and are more randomly scattered in the rolled plate unlike laminations, which tend to be

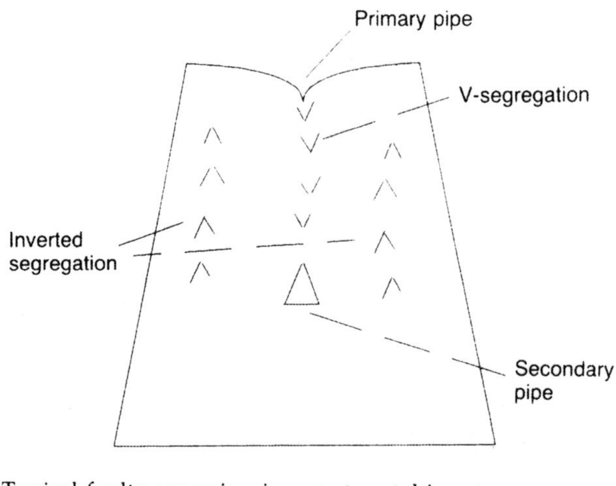

Fig. 7.16 Typical faults occurring in a cast metal ingot.

located on the centre line. As to be expected, these faults occur at one end of a plate or bar, corresponding to the top position of the ingot as originally cast. Variations in grain size do not usually pose so great a problem with forged and rolled products as they do with castings although variations in the directions of rolling can be detected and sometimes they may be significant.

Large forgings such as turbine shafts, which may weigh many tonnes, have the identifications 'top' and 'bottom' at the appropriate ends as determined by the orientation of the original ingot. The top end may thus contain faults and to be potentially weaker than the bottom end which is expected to be 'clean' and therefore destined to become the 'driven end'.

Forgings and rolled plate are tested mainly with normal compression-wave probes either in the single or twin mode. Twin-crystal probes are preferred for testing thin plates and also testing for sub-surface defects in thicker objects. For thicker plates, single-crystal probes are used at frequencies of, say, 4 MHz but reducing to 2 MHz with increase in thickness. Very thin plates of only 1–2 mm thick, known as **sheet metal**, can be tested by means of Lamb waves (section 3.4) which can conveniently be produced by means of a 90° angle probe. Lamb waves can be used to test for laminar faults but, for commercial reasons, this method is not commonly used except, perhaps, at sheet metal mills. The user of sheet metal is more concerned about the ductility of the material and its ability to be drawn rather than the presence of internal laminar imperfections, the testing of which would involve a considerable amount of expense.

Bars and billet sections are either drawn or forged and the reduction may be to such an extent that the central pipe resulting from casting is very fine and may be orientated in any direction. It is thus essential to vary the direction of probing around the bar to ensure that the central fault is detected. With severe drawing in a round bar, there is sometimes a tendency to prooduce a 'burst' (an internal split) which is caused by pressure and lies vertical to the long axis of the bar. This fault can only be detected with an angleprobe search using two probes moving together whilst separated by a skip distance (Fig. 7.17).

When testing a cylindrical bar across its diameter using a compression wave probe placed on the curved surface, problems arise as a result of beam spread which arises from the fact that only a relatively small fraction of the surface of the probe is properly coupled to the surface of the cylinder as a result of the curvature and the source is effectively in the form of a narrow rectangle with the short dimension in the plane of the circular cross-section only about 1 mm in width. This results in a very short near field (section 2.7.1) and, hence a considerable degree of beam spread and the production of echoes due to lateral reflections, in addition to the normal backwall echoes.

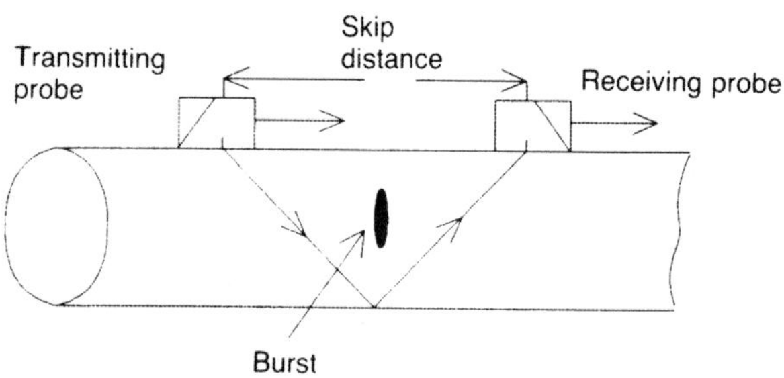

Fig. 7.17 Arrangement of angle probes when testing for 'bursts' in a round bar.

Consider a 50 mm diameter steel cylinder in which the speeds of compression and shear waves are $c_L = 6000\,\mathrm{m\,s^{-1}}$ and $c_T = 3300\,\mathrm{m\,s^{-1}}$, respectively. If the range control of the flaw detector is adjusted to display the first two backwall echoes, it is found that, in addition to the expected ones at 50 and 100 mm, two others appear, at 65 and 84 mm (Fig. 7.18(a)).

Figure 7.18(b) illustrates the circular cross-section of the cylinder, centre O, with the compression wave probe located at A and propagating in the direction AE. The beam spread covers a wide and continuous angle range and, if we consider a ray AB propagated at an angle $\alpha_L = 30°$, reflection takes place at B along the line BCD. Because they are both radii, OA and OB are equal, $r = 25$ mm, so that OAB is an isosceles triangle with the angle OBA also equal to α_L, i.e. 30°, and hence also the angle of reflection OBC. It is clearly seen from the figure that ABC is a right-angled triangle. In a similar manner, reflection takes place at D at the same angle and, from the symmetry of the system, the reflected beam DA must return to the probe at A. ABC is clearly an equilateral triangle with each of its angles 60°. Considering the triangle OBC, BC = $r \cos \alpha_L$ and the total ray path for a complete transit is equal to $6r \cos \alpha_L$, i.e. 130 mm. The flaw detector display is calibrated with respect to a single trip, the screen indication is one half of this quantity, 65 mm.

The 84 mm indication results from mode conversion from compression to shear waves on reflection, whereby part of the reflected beam emerges as shear waves and the remainder as compression waves (sections 2.5.2, 5.3.2). This takes place at B for a suitable value of the angle α_L (Fig. 7.18(c)). As before, ABC is an isosceles triangle so that the angle ABO is also α_L. The angle α_T of reflection, for the shear wave component, is given by Snell's law (equation 2.48):

$$\sin \alpha_T / \sin \alpha_L = c_T / c_L,$$

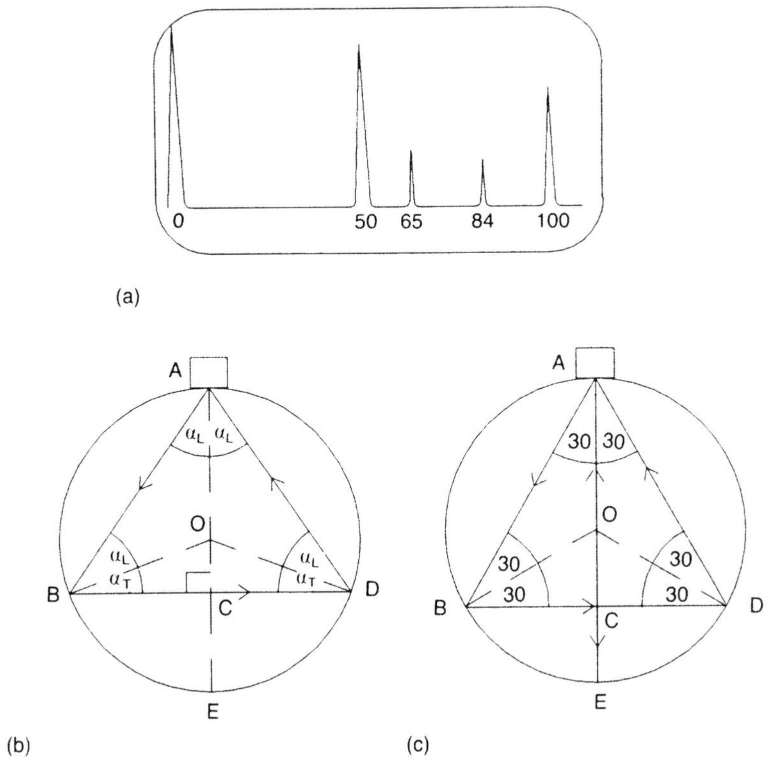

Fig. 7.18 Mode conversion obtained as a result of a compression-wave probe located at a point A on a metal cylinder of 50 mm diameter. (a) Screen display indicating echoes at apparent distances of 50, 65, 84 and 100 mm. (b) Mode conversion at B for compression wave angle of incidence of α_L and shear wave angle of reflection α_T. (c) Reflection without mode conversion for a compression wave angle of incidence of 30° at B.

where c_T and c_L are respectively the shear and compression wave velocities in the material. The reflected shear wave beam then follows the path BD to point D where reflection with mode conversion again takes place and the compression wave component follows the path DA back to the probe at A. This can only take place if the system of rays is symmetrical, for which the triangle ABC must be an isosceles. This arises only for a given value of the angle α_L for which it can be shown (Appendix E) that:

$$\sin \alpha_L = \tfrac{1}{4}\{[(c_T/c_L)^2 + 8]^{1/2} - c_T/c_L\}.$$

Substituting the values of c_L and c_T given earlier gives $\sin \alpha_L = 0.591$ from which $\alpha_L = 36.2°$. From the geometry of the system it is seen that:

$$AB = AD = 2r\cos\alpha_L$$

and
$$BD = 2r\cos\alpha_T,$$

where the radius $r = OA = OB = OD$. The total transit time t for a single echo along the path length ABDA is then given by:

$$t = 4r(\cos\alpha_L)/c_L + 2r(\cos\alpha_T)/c_T.$$

Because the flaw detector is calibrated with respect to compression waves, the effective length of a single trip is given by the product $c_L t$ and, as before, the screen indication is equal to half this quantity, i.e. 84 mm.

Mode conversions from compression to shear or surface waves can also occur in elongated specimens with parallel surfaces (Figs. 5.3(a) and (b)) as, for example, when a probe is placed at position G on the surface of an IIW (A2) calibration block (Fig. 7.1). Reflection takes place at the point P on the opposite surface at a distance of 100 mm and the range is thus set at a distance of 200 mm or more (section 6.2.3). Because of far-field beam spread, reflections take place at the lateral surfaces, which are 25 mm apart. Mode conversion takes place on reflection, with the result that multiple echoes are observed in addition to the back-wall echoes (section 5.3.2). This phenomenon is characteristic of objects such as shafts, axles and rotors when tested from their end surfaces (section 9.4).

7.5 IMMERSION TESTING TECHNIQUES

With the more conventional contact method of ultrasonic scanning, the probe is virtually in contact with the test object and acoustic matching is achieved by means of a thin film of either liquid or grease. With immersion testing (section 5.6), there is a perceivable gap between the probe and object which are usually both located in a tank containing a liquid, mainly water, to maintain acoustic coupling (Fig. 7.19). The scan made in these circumstances is called **gap scan**. Sometimes coupling is achieved by means of a jet of water running to waste, but tanks are normally used and their sizes can range from that of one used for keeping goldfish to that of a swimming pool, depending on the nature of the object under test (Fig. 7.20). Immersion probes, which are waterproof, are invariably of the compression-wave type; they are mostly 10 mm in diameter and operate over a frequency range of 1–25 MHz. The beam from the probe can be focused to a point using a spherical lens and to a line by means of a cylindrical (barrel-shaped) lens (section 2.10). Shear waves of any angle can be produced in the object simply by rotating the probe in its holder by the required amount; thus the number of probes required is very much reduced.

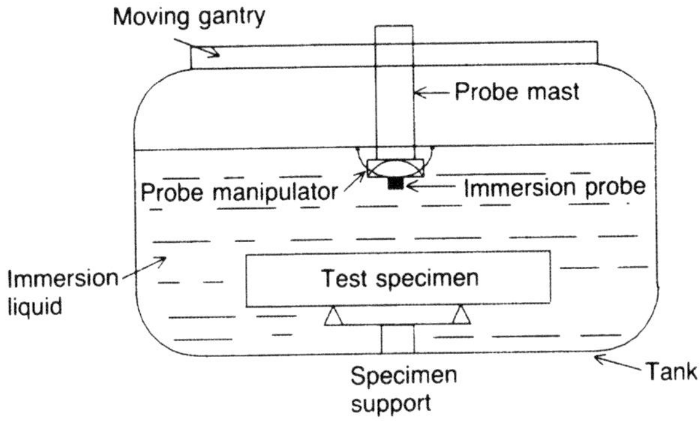

Fig. 7.19 Typical immersion testing arrangement. (Reproduced by courtesy of Krautkrämer GmbH & Co.)

(a)

Fig. 7.20 (a) Immersion testing of turbine disc. Reproduced by courtesy of Krautkrämer GmbH & Co. (b) Very large immersion tank used for testing plates.

(b)

Fig. 7.20 Continued

To achieve satisfactory results, a C-scan plotter should be used. An essential adjunct to the tank (Fig. 7.19) is the carriage which can move in a horizontal plane. A rod, called a 'mast' and a probe manipulator are fitted to the carriage and allow the probe to be rotated and moved vertically. Special attention should be paid to the use of equipment capable of a high degree of both accuracy and resolution in view of the difficulties which may otherwise be encountered in achieving repeatable results, because of the need to control the scanning speed.

In section 5.3.4, it is made clear that the gap between probe surface and the top surface of the material under test, which acts as a delay line, is sufficiently large to prevent interference between echoes from the top and bottom surfaces of the material. Because the speed of compression waves in steel is about four times that in water, this can be achieved by ensuring that the thickness of the water gap is greater than one quarter of the thickness of the metal.

For compression wave testing, it is essential that the probe is positioned so that its axis is accurately perpendicular to the metal surface. For shear wave testing the compression wave probe has to be swivelled from its vertical position to that for which the angle of incidence produces the required angle of refraction, in accordance with equation 2.48: we have an angle probe with the water replacing the

Perspex shoe. Assuming a speed of compression waves in water to be $1483\,\mathrm{m\,s^{-1}}$ and of shear waves in a steel sample to be $3240\,\mathrm{m\,s^{-1}}$, this equation indicates that an angle of incidence of 25.5° gives rise to and angle of refraction of 71.1° for shear waves in the steel. For an adequate degree of accuracy, the probe manipulator should be capable of setting these angles to fractions of a degree.

To achieve a complete coverage of an area of a metal surface, the carriage manipulator must be motorized and be adjustable to provide a raster pattern, either fine or coarse, and all parts of the manipulator, motor and carriage be protected from corrosion. The water in the tank must be treated with a wetting agent and suitable chemicals to guard against undesirable growths such as algae. Furthermore, in some working environments, the use of a small heater may be necessary to stop the water cooling sufficiently to give rise to bubbles on the metal surfaces. The resulting advantages gained from these procedures are:

- the maintenance of a constant degree of coupling;
- independence of operator handling;
- repeatability of results;
- ability to test at higher frequencies as a result of no pressure being exerted on the resultant thinner transducer crystals; and
- the ease of automation.

The last of these advantages can be achieved by fixing an arm to the probe manipulator at the end of which a sensitive pen can be activated by means of a gating device. In this way a 'life-size' C-scan (section 5.6) can be produced on paper. On larger objects such as plates, water jets running to waste are used; here the scan can be speeded up by use of a number of identical probes located at regular intervals along a line perpendicular to the direction of motion. The output from each of these probes, as the plate moves through the jets, is fed into a printer

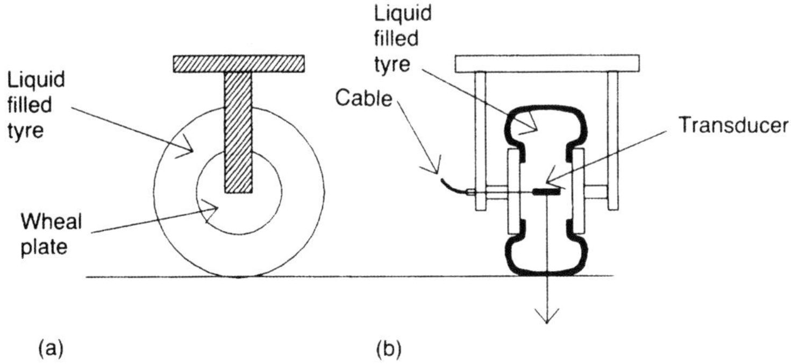

Fig. 7.21 Typical wheel probe. (a) Front view and (b) side view. Reproduced by courtesy of Krautkrämer GmbH & Co.

so that a record in the form of the C-scan is created. Where possible, the pulse-echo technique is used but, with thicker plates, the use of the through-transmission method may be necessary.

Another form of gap scanner is the wheel probe (Fig. 7.21) in which the probe is attached to the axle of a wheel and enclosed in a tyre, the inside of which is filled with oil as a couplant. The tyre is made from a material similar to that used for probe covers. The surface of the test piece is usually sprayed with water and the assembly wheeled over the surface.

REFERENCES

ASTM E317-85 (1985) *Standard Practice for Evaluating Performance Characteristics of Ultrasonic Pulse-Echo Testing Equipment without the Use of Electronic Measurement Instruments*. American Society for Testing Materials, Philadelphia.

BS 1134 (1988) *Assessment of Surface Texture*. British Standards Institute, London.

BS 2704 (1978) (reconfirmed 1983) *Specifications for Calibration Blocks for use in Ultrasonic Flaw Detection*. British Standards Institute, London.

BS 4331 (Part 1) (1978) (reconfirmed 1989) *Assessing the Performance Characteristics of Ultrasonic Flaw Detection Equipment (Overall Performance: On-Site Methods)*. British Standards Institute, London.

BS 4331 (Part 2) (1982) *Methods for Assessing the Performance of Ultrasonic Flaw Detection Equipment (Electrical Performance)*. British Standards Institute, London.

BS 4331 (Part 3) (1974) (reconfirmed 1987) *Guidance for the Internal Monitoring of Probes (excluding Immersion Probes)*. British Standards Institute, London.

DIN 54120 (1966, 1973) *Zerstorungsfreie Prüfung: Kontrollkorper 1*. Deutscher Institut für Normungev, Berlin. N.B English translation available from the British Standards Institute.

ISO 2400 (1972) *Welds in Steel: Reference Block for the Calibration for Ultrasonic Examination*. International Organization for Standardization, Geneva.

ISO 7963 (1985) *Specification for Calibration Block No. 2 for Ultrasonic Examination*. International Organization for Standardization, Geneva.

Krautkrämer, J. and Krautkrämer, H. (1990) *Ultrasonic Methods of Testing Materials*, 4th edn. Springer–Verlag, Berlin, pp 96–102.

Taylor, J.L. (1988) *Basic Metallurgy for NDT*, revised edn. British Institute of Non-Destructive Testing, Northampton.

CHAPTER 8

Flaw sizing in metals

8.1 INTRODUCTION

8.1.1 General considerations

It is often necessary in non-destructive testing to discriminate between the nature of imperfections in materials and it has been stated (API 2X 1988) that the term **flaw** be used to indicate an undesirable condition which may not necessarily require rejection whereas the term **defect** indicates that the fault is sufficiently bad for the object to be declared unfit for use. A number of methods of sizing flaws have evolved over the years, each serving a particular need. Those currently in use are described in this Chapter.

The principal sizing methods are the **probe movement** and **amplitude**, described in sections 8.2 and 8.3, respectively. Small flaws may be sized by either method. If flaws extend beyond the confines of the beam in any direction, then a probe movement method will be required. It is important, when sizing flaws, to take into account the fact that the beam is convergent in the near field and divergent in the far field and to make the necessary correction for the latter (section 2.7).

Flaws may be classified as being either **large** or **small** depending on whether they are larger or smaller than the cross-sectional dimension of the ultrasonic beam at the position of incidence. Sometimes a flaw my be 'large' in one direction and 'small' in a direction at right angles. Examples include a slag line and an artificial target such as the horizontally bored hole used for the 20 dB drop method (section 8.2.2).

8.1.2 Use of the gain control

For accurate flaw sizing, the use of a gain control is essential. This consists of an attenuator calibrated in decibels (section 2.3.6) and located between the amplifier of the flaw detector and the input to the Y-plates of the oscilloscope (Fig. 5.1) so as to permit adjustment of the total gain by the required amounts, usually to the nearest 2 dB but with some digitized equipment it can be as little as 0.1 dB. It is sometimes known as the decibel control.

At the time of writing, because of a lack of a suitable international agreement, there is no common practice among ultrasonic instrument manufacturers about how the decibel units are displayed: for an increase in signal amplitude (as displayed on the screen) with some instruments the decible values for gain increase but, with others, they decrease. This should present little problem if the instruments used 'in house' are all from the same manufacturer but caution should be exercised if other equipment is introduced. Thus the use of a positive or negative sign given to a decibel indication can mean either an increase or decrease with respect to a standard value so that, in the absence of any international agreement, consistency in interpretation is essential. Table 8.1 illustrates examples of the use of positive and negative decibel values as provided by BS 3923 Part 1 (1986) and IIS/IIW/850/86 (1987). They appear to be contradictory, but the former relates to gain settings and the latter to a reject level (Simpson 1989).

8.2 PROBE MOVEMENT SIZING METHODS

8.2.1 The 6 dB drop or half-beam technique

The 6 dB drop or half-beam technique provides the simplest means of sizing large planar flaws such as laminations (Fig. 8.1). The following procedure is adopted.

Table 8.1 Alternative uses of decibel scales for expressing sensitivities
(a) Minimum scanning acceptance levels from British Standard BS 3923 Part 1 (1986)

Examination level	*1*	*2*	*3*
Normal beam scans	DAC + 14 dB	DAC + 8 dB	DAC + 8 dB
Scans for transverse imperfection	DAC + 20 dB	DAC + 14 dB	DAC + 8 dB

Here DAC + dB indicates an **increase** in sensitivity
DAC = distance amplitude correction (section 8.3)

(b) Recommended acceptance/rejection criteria from IIW/850/86 (1987)

Class	*1*	*2*	*3*
Recommended thickness (mm)	7–50	7–100	7–300
Evaluation threshold	DAC − 10 dB (33 % DAC)	DAC − 14 dB (20 % DAC)	DAC − 14 dB (20 % DAC)
Amplitude rejection level	DAC (100 % DAC)	DAC − 4 dB (65 % DAC)	DAC − 4 dB (65 % DAC)

Here DAC − dB indicates and **increase** in sensitivity

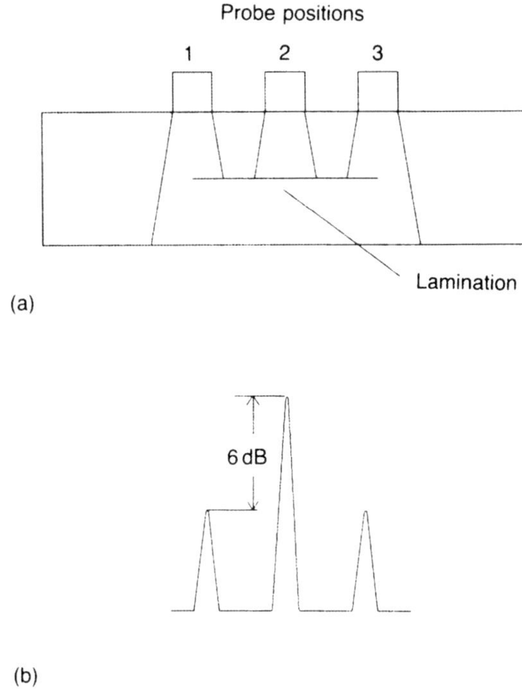

Fig. 8.1 Measurement of the length of a lamination in a plate using the 6 dB drop method showing: (a) the positions of the probe, 2 for maximum output and 1 and 3 for 6 dB below the maximum; and (b) the corresponding echo signals.

1. Locate the flaw;
2. maximize the echo from a central part of the flaw;
3. move the probe around until the flaw echo begins to decrease in amplitude;
4. stop when the amplitude of the echo has been reduced by 6 dB, (one half) for which the centre of the probe is located over the edge of the flaw;
5. mark the position of the probe; and
6. repeat the procedures (3.), (4.) and (5.) as necessary, so as to obtain a plot of the shape of the defect.

A variation of the 6 dB drop technique, which is used for testing plates, involves adjusting the decreasing amplitude of the echo from the flaw to equal the increasing amplitude of the echo from the back wall of the plate. This method provides results similar to those obtained from the basic technique. Greater accuracy is obtained with the use of a single probe than with a twin or double probe because the former has a more symmetrical beam shape than the latter. With thinner plates, for which the use of single probes may be inconvenient because of dead-zone problems (section 5.3.1) and the consequent requirement for the use of twin probes, the degree of accuracy may be somewhat reduced. In the latter instance it clearly cannot be assumed, as when using single probes, that the centre of the probe coincides with the edge of the flaw at the position at which the amplitude has dropped by 6 dB. Plate-testing standards such as British Standard BS 5996 (1980) usually provide guidance to the thickness limits relating to the uses of both twin and single probes together with the frequency range to be used. For example, twin probes are recommended for plate thicknesses of 5–100 mm and single probes for thicknesses of 20 mm and over. The length of the dead zone should not exceed 15 % of the plate thickness or 15 mm, whickever is the lower.

8.2.2 The 20 dB Drop Method

The 20 dB drop method is the one most often used to size small flaws by means of the probe-movement method and is very much favoured for testing welds with angle probes. The following procedure is adopted.

1. Initially obtain a plot of the beam profile with the use of a beam calibration block such as the BS 2704 A5 (section 7.2.2, Fig. 7.6) containing several holes at different depths.
2. Locate each hole and maximize the amplitude of the echo so as to be satisfied that the centre of the beam is striking the hole.
3. Adjust the echo signal to full screen height and note the depth of the hole and the position of the probe.
4. Move the probe forward until the amplitude of the echo falls by 20 dB, again noting the position of the probe and its distance from the hole.
5. Repeat (4.) with the probe moving in the opposite direction.
6. Repeat (3.), (4.) and (5.) for the other holes.

Sufficient information is now available to produce a beam profile plot (Fig. 8.2) which is usually done on specially prepared cards such as those supplied by the Welding Institute (UK) (Appendix G), although it is possible to use ordinary graph paper for this purpose. If profiles are being prepared for work with angle probes, a separate one is required for each of these probes. Because this work is laborious, it is advisable to preserve the diagram for future use provided it can be confirmed that the characteristics of the probe have not changed.

The following procedure is then adopted for sizing the flaws.

1. Locate the flaw and maximize the echo signal.
2. Note the position of the probe and use the screen indication to obtain its distance from the flaw.

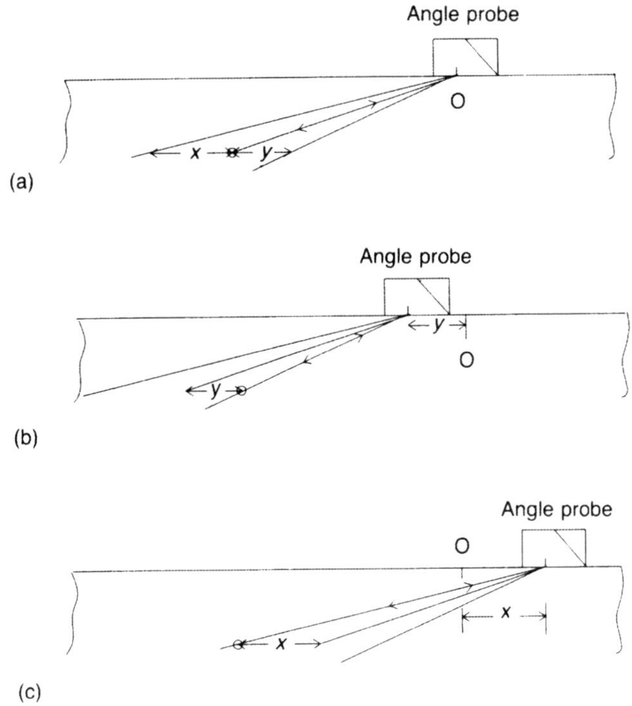

Fig. 8.2 Use of an A5 or similar block for assessment of the beam width of an angle probe using the 6 dB drop method showing: (a) the probe at position O for maximum echo; and (b) and (c) the probe moved, respectively by distances y to the left and x to the right of O to receive echoes of amplitude 6 dB below that of the maximum.

3. Move the probe forward and then backward noting the 20 dB drop positions, as before, and record the probe positions and the screen readings (three pairs of readings in centre, forward and backward positions are recorded).
4. Use the readings, with the aid of the beam profile and a cross-sectional view of the object under test, to construct the position and size of the flaw. This exercise is made easier with the use of a plotting device (Fig. 8.3) available from the Welding Institute (UK) and other sources.

8.2.3 The time-of-fight diffraction method

The time-of-flight diffraction method (TOFD) has been used for many years for measuring crack depths in concrete (section 10.2.7) but more recently Silk (1977) has extended the method for sizing both surface and sub-surface cracks in metals and these techniques are now

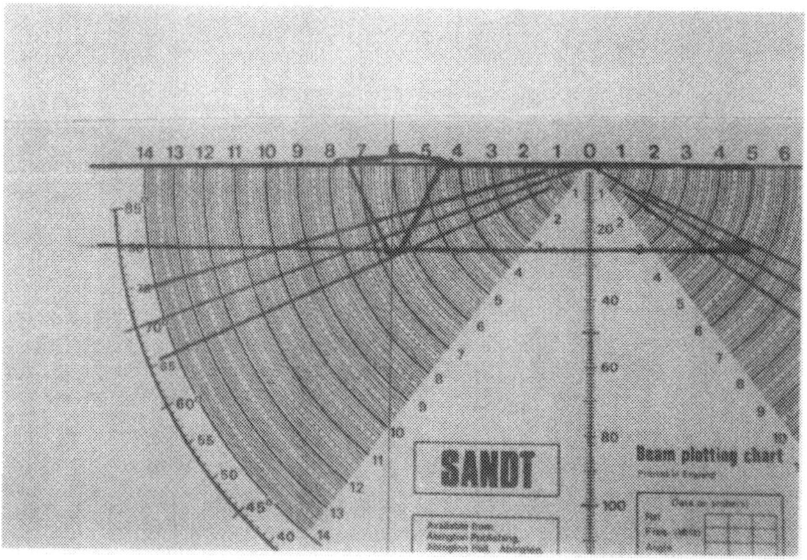

Fig. 8.3 Flaw location ruler used for beam profile plots. Reproduced by courtesy of Woodhead Publications at TWI.

extensively used. As with crack sizing in concrete, the method is based on the phenomenon of diffraction at edges (section 2.7). Figure 8.4 illustrates a deep crack of length *l*, orientated perpendicular to a plane surface of an object with its upper extremity of the crack at a depth *d*. Compression wave megahertz frequency angle probes having angles of 40°–60° (section 4.4) and having wide beam spreads are normally used for this purpose. The pulse transmission method (section 5.4) is most commonly employed with the two probes located at a skip distance apart (section 7.3) on opposite sides of the flaw.

The effect of a wide beam spread is that surface transmission of

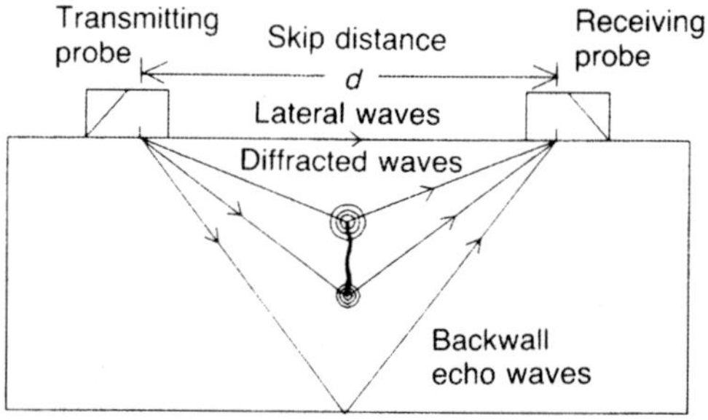

Fig. 8.4 Use of time-of-flight diffraction for crack measurement.

compression waves, sometimes known as lateral or creeping waves, takes place in the same way as experienced with the testing of concrete (section 10.2.6) and results in waves propagated directly from the transmitting probe to the receiving probe, in addition to the backwall echo (Fig. 8.4).

Figure 8.5 illustrates the corresponding echoes A and C and also echoes B1 and B2 resulting from diffraction at the extremities of a crack, as they would appear on the oscilloscope screen. With only longitudinal wave propagation, the speed of travel is uniform, with the result that no complications arise with calibration, as experienced with shear waves which are rarely used for this technique.

By adjusting the positions of the probes to provide a minimum value of time of flight, the crack tip is then located exactly midway between the probes. For a crack orientated in a direction vertical to the surface, the size and depth can be evaluated in a manner similar to that used in deriving equation 10.6. Referring to Fig. 10.7 we have that d is the depth of the upper crack tip and l is the length of the crack. Applying Pythagoras's theorem we have:

$$d^2 + s^2 = (ct_1/2)^2,$$

and

$$(d + l)^2 + s^2 = (ct_2/2)^2,$$

where c is the speed of compression waves and t_1 and t_2 are the respective time delays for the ultrasound diffracted at the upper and lower tips of the crack. Solving these simultaneous equations gives:

$$d = [(ct_1/2)^2 - s^2]^{1/2}, \tag{8.1a}$$

and

$$l = [(ct_2/2)^2 - s^2]^{1/2} - d. \tag{8.1b}$$

Errors in estimating the depth of the crack tip increase as the depth decreases. When using an unrectified A-scan display (Fig. 8.5) it will be seen that crack-tip echoes together with the lateral-wave and backwall echoes suffer phase reversals. Thus the signal from the lower tip of the

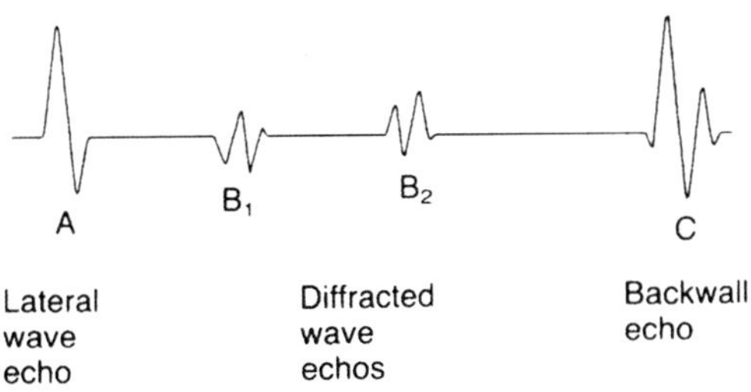

Fig. 8.5 A-scan display of signals from time-of-flight diffraction measurement as (illustrated in Fig. 8.4).

flaw is in phase with the lateral wave with the result that one can easily identify the crack-tip signals.

Much effort has been expended on the production of computer software for the collection, storage and analysis of ultrasonic data. One example is known as ZIPSCAN (developed at the National Nondestructive Testing Centre at Harwell (Appendix G). It provides real time displays in the forms of A-, B-, C- and D-scans. The D scan is the name given to a longitudinal scan obtained by moving, along the surface, two probes maintained at a fixed distance apart, as indicated in Fig. 8.6, in a direction at right angles to the line between them so as to obtain a profile of the flaw in this direction. Originally it was the responsibility of the operator to identify flaw signals appearing on the oscilloscope screen and to list the position, size and type of flaw but, nowadays, computer software is available for this purpose. It has been found, after an extensive number of trials, that the time of flight diffraction method has produced consistently accurate data. A British Standard (BS 7706, 1993) relating to this subject is now available.

8.3 AMPLITUDE METHODS

The mostly used methods of defect sizing are the uses of the comparator block, the distance amplitude correction (DAC) method and the distance gain size (DGS) techniques. They are discussed in the following sub-sections.

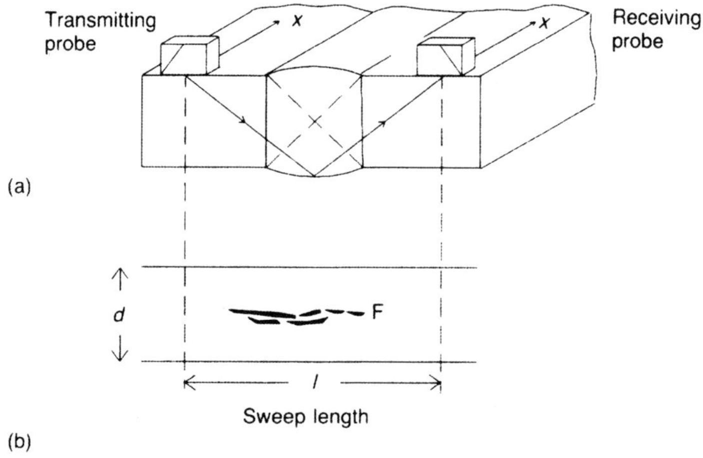

Fig. 8.6 D- or longitudinal scan in x-direction along the length of a weld in a sample of thickness d, showing: (a) arrangement of probes at constant separation; and (b) the flaw indication.

8.3.1 The comparator block

The use of a comparator block provides the most straightforward amplitude method of sizing in that it consists simply of comparing the echo from a flaw with that from an artificial target such as a flat-bottom hole in the block. If it is made from the same material as that of the workpiece and has a similar shape and surface finish, with the target located at a depth similar to that of the flaw, the comparator block provides a simple way of estimating a flaw size. Comparator blocks can also be used to set sensitivity levels for nominal flaw sizes at given depths but, if one has a wide range of defect sizes and depths to consider, a large variation of these blocks would be necessary and may prove to be highly expensive.

A typical block (Fig. 8.7) would have a thickness of 150 mm and a diameter of 75 mm but with smaller thickness for shallower holes. The blocks should be sufficiently wide to avoid contact with spreading beams. The holes are drilled from the bottom of the block and usually plugged to prevent the ingress of oil and water. Comparator blocks are almost invariably used with normal probes.

8.3.2 The distance amplitude correction method

Because of beam spread, the echo heights observed on the screen of a flaw detector decrease with distance in the far field (section 2.7.1). Bearing in mind that the distance in the far field from the source varies inversely with wavelength, this phenomenon is particularly noticeable for shear waves which have wavelengths about twice those for the corresponding compression waves in a given material. A decay curve indicating far-field effects can be drawn on a graticule placed over the

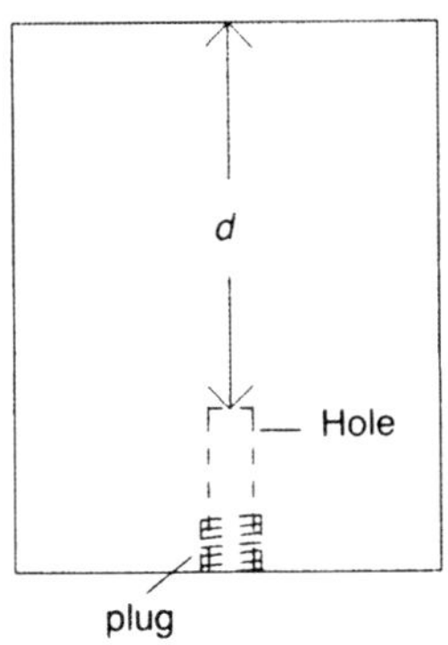

Fig. 8.7 Cylindrical comparator block with flat-bottom hole drilled in lower surface.

oscilloscope screen; this is known as the DAC curve, which indicates that a small echo signal at a remote distance from the probe may have a similar significance as a larger one closer to the probe.

With normal probes, blocks similar to the comparator blocks may be used to create the curve but, with angle probes, horizontally side-drilled holes are used. A typical example is the BS 2704 A5 block (section 7.2.2, Fig. 7.6). Where possible, the use of side-drilled holes in the workpiece may be feasible. For some standards, the diameters of the holes can be varied by amounts compatible with the thickness of the workpiece. The amplitude of an echo from a reflector depends on the nature of the latter. For disc reflectors, which includes flat-bottom holes, it can be shown (Krautkrämer and Krautkrämer 1990) that the amplitude a of a reflected pulse is proportional to the area and thus to the square of the diameter D:

$$a_2/a_1 = (D_2/D_1)^2. \tag{8.2a}$$

Expressed in decibels this becomes:

$$\text{increase in level} = 20\log(D_2/D_1)^2 = 40\log(D_2/D_1). \tag{8.2b}$$

Thus a doubling in diameter would result in an increase of 12 dB.

The amplitude of an echo also depends on the distance, d, of the disc from the source. Thus if the value of d is increased from d_1 to d_2, for a given diameter, D, the echo amplitude reduces and the ratio becomes:

$$a_2/d_1 = (d_1/d_2)^2. \tag{8.3a}$$

Expressed in decibels, we have:

$$\text{increase in level} = -40\log(d_1/d_2). \tag{8.3b}$$

Thus a doubling in distance would result in an increase of −12 dB: a **decrease** of 12 dB.

However, a side-drilled hole acts at a cylindrical reflector, for which it can be shown that the increase in amplitude, a, at a fixed distance, with diameter D is given by:

$$a_2/a_1 = (D_2/D_1)^{1/2}. \tag{8.4 a}$$

Expressed in decibels, this becomes:

$$\text{increase in level} = 10\log(D_2/D_1). \tag{8.4 b}$$

Thus a doubling in diameter would result in an increase of only 3 dB.

As before, the amplitude of an echo also depends on the distance, d, of the disc from the source and, for an increase in distance, d_1 and d_2, for a given diameter, D, the echo amplitude again reduces and the ratio, this time, becomes:

$$a_2/a_1 = (d_1/d_2)^{3/2}. \tag{8.5 a}$$

Expressed in decibels, we have:

$$\text{increase in level} = -30\log(d_1/d_2). \qquad (8.5\text{ b})$$

Thus a doubling in distance would result in a **decrease** of 9 dB.

To produce a DAC curve on the screen of the oscilloscope from calibration blocks with either flat-bottomed holes, for use with normal probes, or side-drilled holes for use with single probes, the following procedure should be adopted.

1. Consider the requirement that the holes should be located at a variety of depths within the expected range of working;
2. choose a hole close to the surface in a position to give the highest echo amplitude and adjust to, say, 80 % FSH;
3. mark this position on the screen with a suitable pen or crayon;
4. locate the other holes and mark the maximized peak position on the screen;
5. join up the marks when a sufficient number has been made to produce a 100 % DAC curve; and
6. draw other curves for 50 % or 20 % values below the 100 % curve, i.e. 6 or 14 dB lower, to help with the defect evaluation (Fig. 8.8).

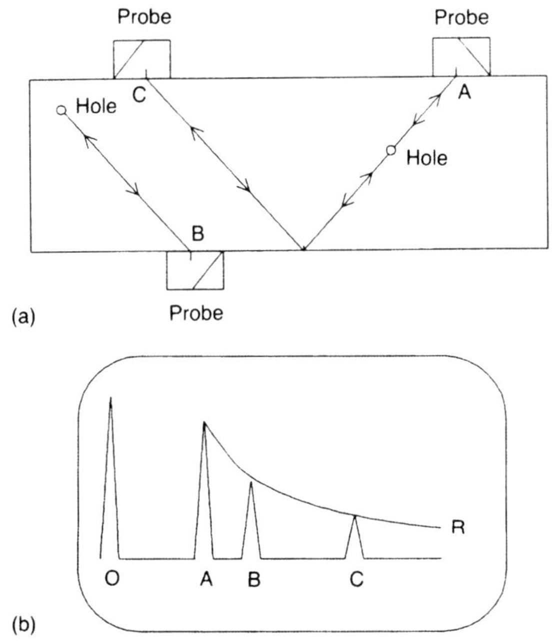

Fig. 8.8 (a) Use of A5 or similar calibration block indicating angle-probes positions A, B and C to obtain echoes from the holes and (b) corresponding echo signals producing a primary reference level (i.e. 100 % DAC) curve, R, drawn through the peaks of echoes A, B and C. O = transmission peak.

To avoid repetition, it is sometimes the practice to draw the curves on clear plastic film and to replace them whenever different probes are used. A similar procedure can be adopted with DGS scales (section 8.3.3).

8.3.3 The distance gain size method

The distance gain size (DGS) method was originated by J. Krautkrämer (1959) under the German title of *Abstand Verstarkung Grosse (AVG)*. It is a graphical representation of the various echo magnitudes produced by different sizes of reflectors located at different distances from a probe. It shows clearly the effects of the locations of small reflectors in the near field and the exponential decay of amplitude in the far field (section 2.7.1). The x-axis of the graph, representing distance, is graduated in multiples of near-field length and the y-axis indicates gain in decibels. Curves are plotted for different sizes of reflectors, in the form of discs, expressed in multiples of the diameter of the probe. This single graph can thus be used to indicate the size of any reflector smaller than the cross-sectional area of the beam at any distance from the source if equating the size of a disc-shaped target to that of a flaw and thus eliminating any kind of comparator block (Fig. 8.7).

Later DGS diagrams were produced for use with specific probes, one for 10 mm diameter and another for 20 mm diameter. but now DGS diagrams are customized for individual probes and types of metal test pieces. An advantage of customization is that dimensions in both millimetres and inches can be provided. These diagrams may not give the true sizes of flaws but they relate the flaw size to an equivalent disc size exactly in the same way as with a comparator block, as an equivalent flaw size (EFS). It never happens that the actual size is smaller than the estimated size because of the existence of irregularities in shape and surface finish of the former. DGS diagrams and comparator blocks can also be used for setting sensitivity levels when conducting tests.

The use of the DGS diagram (Fig. 8.9) for compression wave probes enables sensitivity settings, available gain and flaw size estimates to be made with the following procedure.

1. Choose the probe to be used and the corresponding diagram.
2. Measure the thickness of the component (e.g. 400 mm).
3. Mark the position of the back wall along the distance axis of the diagram and draw a vertical line from this point until it intersects the backwall echo line. Anything observed to the right of this line can be disregarded.
4. Measure the depth of the flaw if not already known (e.g. 150 mm).
5. Mark the position of the flaw depth on the distance axis and draw a vertical line through it.

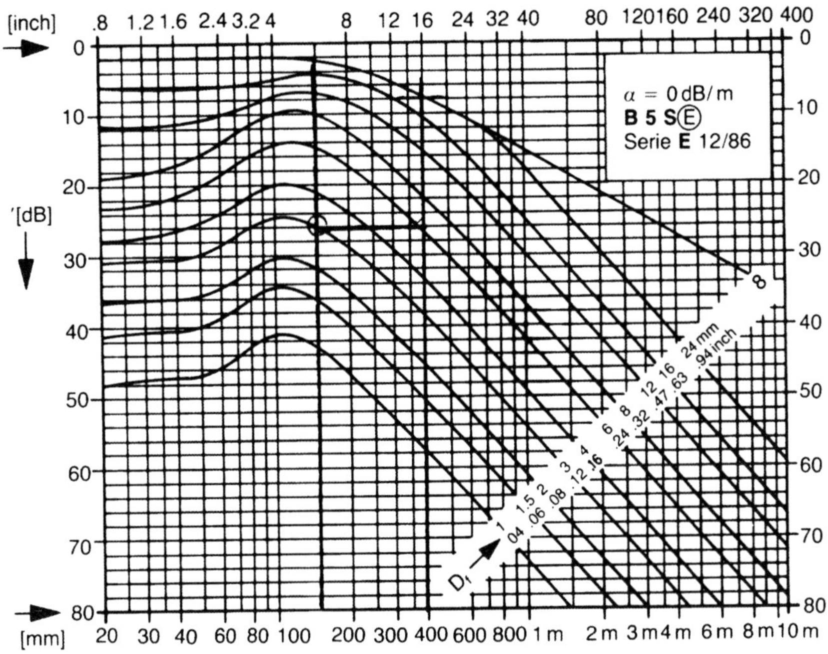

Fig. 8.9 DGS diagram for a 24 mm, 5 MHz compression-wave probe. (Reproduced by courtesy of Krautkrämer GmbH & Co.)

6. Measure the difference, in decibels, between the flaw echo and the backwall echo, e.g. −18 dB. Although it is usual for the flaw echo to be smaller than the backwall echo, occasionally it may be slightly larger and a positive decibel reading would be obtained.
7. From the point on the diagram where the backwall line echo line was intersected, measure downwards in decibels (upwards if the decibel reading is positive). Draw a horizontal line from this point to the left until the flaw line, as obtained in (5.), is intersected. The flaw size can now be estimated from the position of this point as indicated by the size curves (e.g. 3 mm).

If the attenuation coefficient α of the material (sections 2.4.2 and 8.4) is known, a correction factor can be introduced to the estimated flaw size by calculating the difference in distance between the flaw echo and the backwall echo and multiplying it by α so to obtain the attenuation in decibels. The latter is then subtracted from the estimated value, f, to provide a corrected value. For relatively thin samples or when the flaw is very close to the back wall, this correction may be sufficiently small so as to be ignored.

The following example of the use of DGS diagrams shows how it is possible to determine in advance what would occur if a 4 MHz probe having a diameter of 10 mm is replaced by another of the same frequency but having a diameter of 24 mm when testing a block of steel 100 mm thick with a flaw having an expected depth of 50 mm. If it is assumed that a backwall echo is set to full screen height (FSH) in each case before a search, then the flaw echo height relative to the backwall datum is considerably different, as shown in Figs. 8.10 and 8.11. In each case the flaw size is 3 mm; the echo is reduced by 4 dB for the 10 mm probe but by 22 dB for the 24 mm probe. This is an example of what can occur with operators at different sites and, for the same specimen, resulting in the use of the 10 mm probe giving a large flaw indication and that of the 24 mm probe giving a very small one, thus giving rise to a rejection at one site and an acceptance at the other simply because the probes, of identical frequency, have different diameters.

DGS diagrams are also available for use with angle shear wave probes and the method of use is the same as that for compression wave probes except that the establishment of a back wall is usually obtained

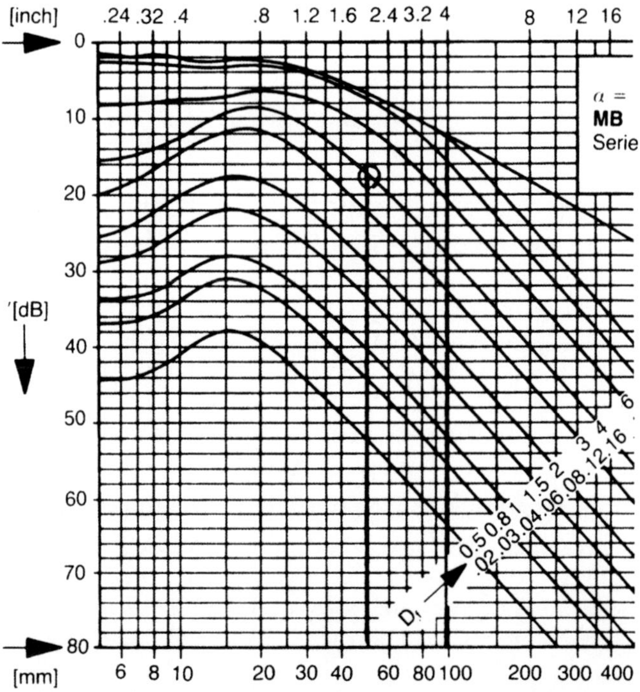

Fig. 8.10 DGS diagram for a 10 mm 4 MHz compression-wave probe showing the decibel relationship between a backwall echo at 100 mm distance and a 4 mm flaw at 50 mm. (Reproduced by courtesy of Krautkrämer GmbH & Co.)

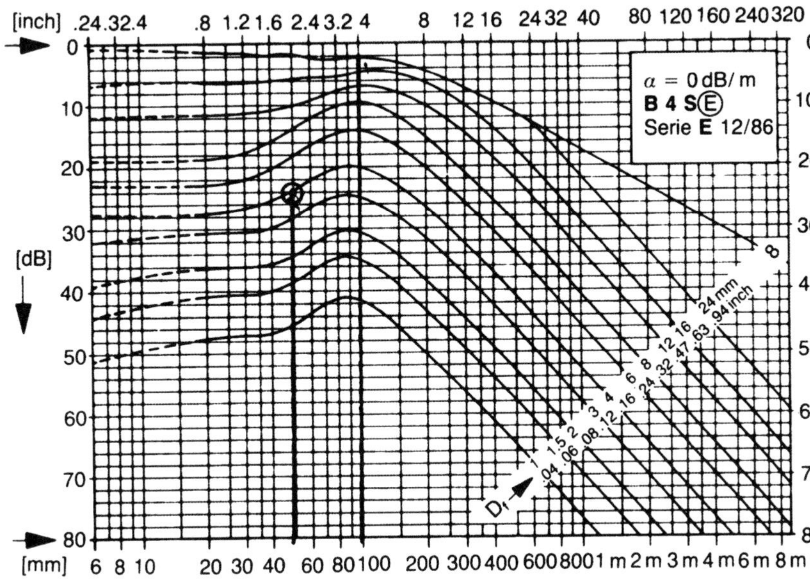

Fig. 8.11 DGS diagram for a 24 mm, 4 MHz compression-wave probe showing the decibel relationship between a backwall echo at 100 mm distance and a 4 mm flaw at 50 mm. (Reproduced by courtesy of Krautkrämer GmbH & Co.)

by means of a calibration block,using either the 25 or 100 mm radius as an artificial target. If, for example, the 100 mm radius is chosen, the vertical line representing the backwall echo is drawn on the chart at the 100 mm position and would intercept the line representing the backwall echo (as indicated in the example described earlier). The distance between the probe and the flaw is measured with an ultrasonic flaw detector and the flaw line can then be drawn (e.g. to the vertical line marked 60 mm). The flaw signal height is then compared with the calibration signal for the 100 mm radius in the same way as with a compression wave probe (Fig. 8.12). It is possible to make a correction for attenuation but the procedure is considerably more complex than with a compression wave probe and should be attempted only if it is highly essential to do so. The reader who is interested in this procedure should consult the appropriate British Standard (BS 3923 Part 1 1986).

DGS scales, derived from DGS charts, are available for compression wave probes and for shear-wave angle probes and are designed to fit in front of the oscilloscope screen. Figure 8.13 illustrates an example of a scale for a 4 MHz 24 mm diameter compression-wave probe. The sensitivity level is adjusted with respect to the backwall echo of the specimen. Equivalent flaw sizes (EFS) are assigned to curves on the scale by adjusting the sensitivity according to the instructions provided

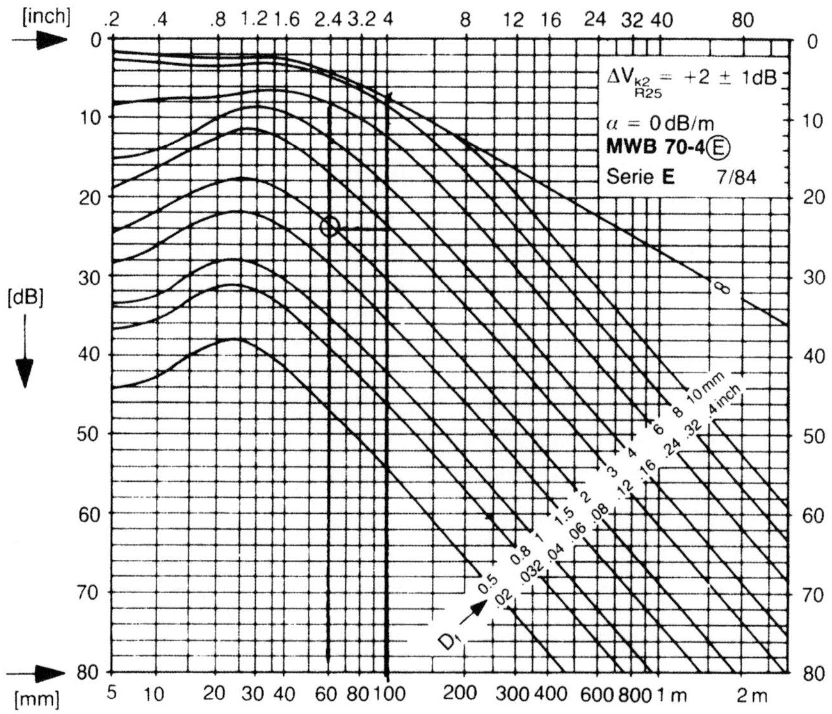

Fig. 8.12 DGS diagram for a 10 mm, 4 MHz 70° angle probe. (Reproduced by courtesy of Krautkrämer GmbH & Co.)

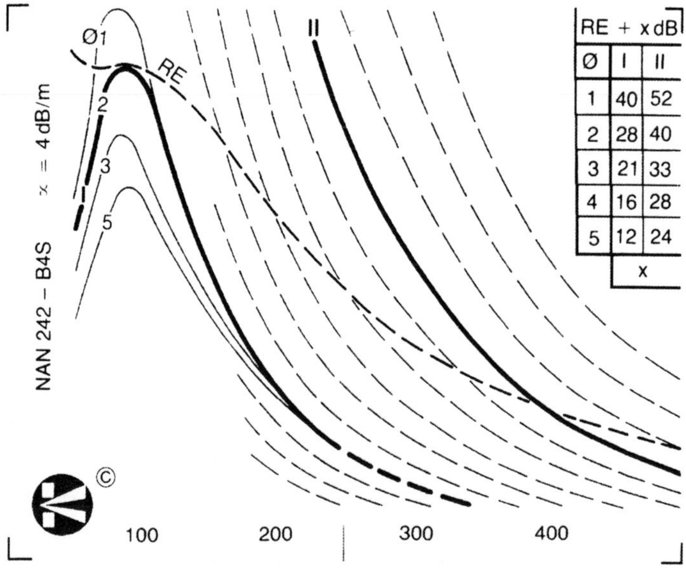

Fig. 8.13 DGS scale for a 24 mm, 4 MHz compression-wave probe. (Reproduced by courtesy of Krautkrämer GmbH & Co.)

for the scale so that the thick line I represents a 2 mm diameter target. The backwall echo is then adjusted and 28 dB is added. Alternatively, the thick line II can be used, for which 40 dB is added. It should be noted that the difference between the thick lines I and II represents 12 dB. Thus, whatever EFS diameter is assigned to line I, the value for line II is double the diameter represented by line I.

8.4 ATTENUATION CORRECTIONS

For flaw sizing in highly attenuating materials corrections may need to be made to allow for any differences between the structure of the material of the specimen under test and that of the calibration block used to establish the sensitivity level. This requires the measurement of the attenuation coefficient of the material (section 5.3.3), which may be expressed in decibels per metre, decibels per millimetre or decibels per inch. It should be borne in mind that, with echo measurements, the distance over which there is a given loss in decibels is equal to the total sound path, twice the depth of the test sample. To avoid complications with the measurements, it is best to use a probe having a small diameter so as to achieve a near-field length sufficiently short for all the readings to be taken in the far field. Allowances must, of course be made for the 6 dB correction for doubling of distance. Care must be taken to avoid the effects of a divergent beam striking the lateral walls of the material under test (Fig. 5.3 (a)).

The following procedure should be adopted for measurements of attenuation.

1. Place the probe on the surface of a flaw-free and parallel section of the material to be tested to obtain a pattern of multiple echoes and select two backwall echoes either in the far field or at a sound-path distance greater than three times the near field. The positions of the selected echoes should be such that the second echo is double the distance from the zero position than that of the first.
2. Measure the difference A' in decibels of the heights of these echoes and note the distance between them.
3. It is necessary to use a probe of diameter and frequency compatible with **both** echoes to be in either the near field or the far field (section 2.7.1). If they are both in the far field, subtract 6 dB from A' to obtain the value of the attenuation loss A. No correction is necessary if they are both in the near field and $A = A'$.
4. Divide the attenuation loss by the path length, d, between the echoes.

Thus the attenuation coefficient (or factor) α is given by:

$$\alpha = A/d.$$

It should be noted that, because of the variation of attenuation with frequency (section 3.3.1), often by large amounts, the value of the frequency should be recorded otherwise the result becomes meaningless.

The test for transverse-wave attenuation can be carried out with a shear-wave transducer (as opposed to a probe), e.g. a Y-cut quartz crystal or a Z-cut ceramic (section 4.3) but difficulties would arise with a conventional type of couplant which is unlikely to support shear stresses. It is possible, in spite of their inability to produce multiple echoes, to use a pair of angle probes in the transmission mode (Fig. 8.14). The following procedure may be adopted.

1. Maximize the received echo with the probes in line and located one skip distance apart.
2. Move one of the probes until the distance of separation is equal to two skip distances and note the decibel drop from the level at the first position.

The decibel loss A is caused by both attenuation and beam spread. If both echoes lie within the far field, the beam-spread loss is equal to 6 dB. It is then necessary to consult a DGS diagram for the angle probes in use to determine the actual loss due to beam spread (Fig. 8.15).

The transverse wave attenuation factor (or coefficient) α_t is obtained from the expression:

$$\alpha_t = (A' - L)/d,$$

where A' is the decibel difference between the first and second skip distances and d is the distance travelled by the ultrasound between the first and second skips. L is the decibel loss obtained from the appropriate DGS diagram. Here again the value of the frequency must be quoted.

When the sensitivity of the equipment is determined by means of either a calibration block, comparator block or reference standard, allowances may have to be made for surface finish, curvature and difference in thickness of the calibration block and the test sample. The total attenuation is given by the sum of the decibel loss in the material and that caused by the above mentioned factors. The difference between the decibel loss in a given test sample from that in the calibration

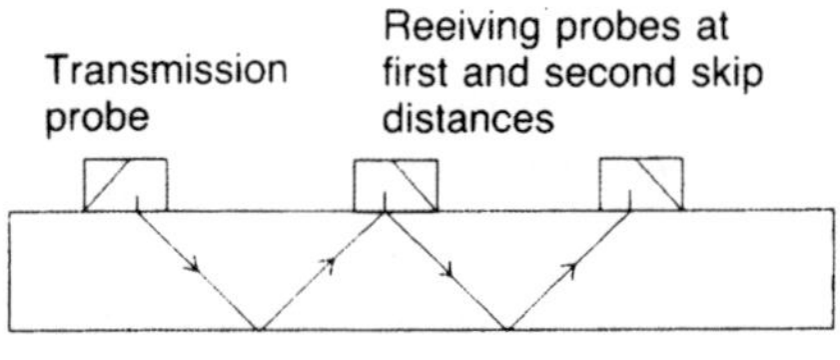

Fig. 8.14 Use of angle probes to measure transverse-wave attenuation.

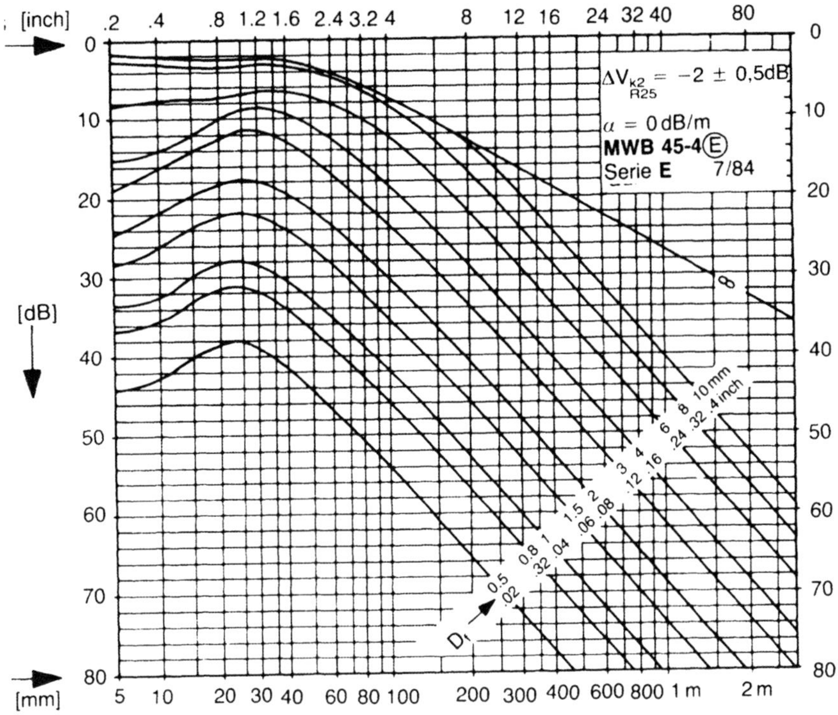

Fig. 8.15 DGS diagram for a 10 mm, 4 MHz 45° angle probe. The backwall echo line is represented by the ∞ (infinity) line.

block is called the **transfer loss**. This is determined with the use of two similar angle probes set at one skip distance apart one the surface of the calibration block and sample, in turn, and the transfer loss is equal to difference between the two attenuation measurements. The measurements of the two attenuations, which is a time-consuming process, is described in the procedure laid down by British Standard BS 3923 Part 1 (1986). Because of the complexities involved, it can often be assumed that the transfer loss correction is simply taken as the difference in decibels between the backwall echo in the calibration block and that in the specimen. In most instances, this assumption does not introduce a significant difference to the result.

REFERENCES

API 2X (1988) *Recommended Practice for the Ultrasonic Examination of Offshore Structural Fabrication*. American Petroleum Institute of Recommended Practices, Washington, DC.

BS 3923 Part 1(1986) *Manual examination of fusion welds in ferritic steels.* British Standards Institute, London.

BS 5996 (1980) *Methods for ultrasonic testing and specifying quality grades of ferritic steel plate.* British Standards Institute, London.

BS 7706 (1993) *Methods for Ultrasonic Time of Flight Diffraction.* British Standards Institute, London.

IIS/IIW/850/86 (1987) *The Evaluation of Ultrasonic Signals.* The Welding Institute, Abington, Cambridge.

Krautkrämer, J. (1959) *Br J Appl Phys*, **10**, 240–5.

Krautkrämer, J. and Krautkrämer, H. (1990) *Ultrasonic Methods of Testing Materials*, 4th edn. Springer, Berlin, Section 5.2.

Simpson, G. (1989) *Br J NDT*, **31**, 33–4.

Silk, M.G. (1977) Chapter 2, in *Research Techniques in Nondestructive Testing*, Vol 3 (ed R.S. Sharpe), Academic Press, London.

CHAPTER 9

The testing of metals

9.1 GENERAL CONSIDERATIONS

This Chapter discusses the ultrasonic testing of a variety of metal products including plates, billets, shafts, rotors, rods, bars, pipes, welds, rivets and fasteners, castings and objects having complex shapes. Although common factors exist with the testing of many of these items, problems which are unique to a particular type of product may arise. These are discussed in the following sections.

9.2 ROLLED PRODUCTS

9.2.1 General considerations

Rolled products include plates, slabs, sheet metal, billet and rolled sections and they are derived from an ingot which is heated and rolled out to produce the intended shape (section 7.4). Any faults existing in the original ingot would be found in the finished product. With plates, slabs and sheet metal, the flattened and elongated flaws are not difficult to detect because they tend to lie in a direction parallel to the test surface but, for rolled sections, the effects of surface contours makes flaw detection more difficult. Plate testing in the UK should be performed in accordance with British Standard BS 5996 (1993) which provides guidance on the types of ultrasonic probes to be used and acceptance levels for internal imperfections. The whole surface area is scanned backwards and forwards and the spacing between successive scans is selected to suit the required acceptance level. The latter provides the choice of seven grades covering spacings ranging from 15 to 100 mm and the minimum tolerable imperfection size depends on the chosen grade. The minimum imperfection size per grade can vary from 5 × 5 mm to 100 × 20 mm.

A flaw is characterized either as an internal imperfection or a defect. The former is regarded as any planar or laminar imperfection and the latter as an unacceptable imperfection exceeding either the maximum size or a certain population density.

Testing is conducted either on a material in the 'as rolled' condition or a material which, after heat treatment, is sufficiently flat to ensure the performance of a valid test. The surfaces must be free from any foreign matter such as dust, rust, scale, grease and paint. For either manual or assisted manual testing, a pulse-echo instrument must be used and, for automatic testing, the gates and monitors should be capable of differentiating between acceptable and unacceptable imperfections. A liquid couplant should be used and it must not have any detrimental effect on the material under test and, furthermore, should be capable of conducting ultrasound both in and out of the material.

Compression wave probes should be used for testing and, for plates of 5–20 mm thick, the use of a twin-crystal probe is to be preferred. For thicknesses of 20–100 mm, either single or twin probes may be used but, above 100 mm thickness, the use of single-crystal probes is necessary. The minimum operating frequency should be 2 MHz unless the material under test is highly attenuating. Here, the use of a lower frequency probe would be necessary. The maximum frequency for manual testing is 5 MHz but, for automatic testing, a frequency of up to 10 MHz is permitted. If the dead zone exceeds 15 % of the thickness, the material should be tested from each side.

For both manual and automatic testing, the surface of the material should be scanned along equally-spaced parallel lines at a speed not greater than 150 mm s^{-1}. For automatic scanning, however, the pulse repetition frequency for each probe should be sufficiently high to ensure that any relevant imperfections are detected. With a plate which is to be welded at any of its edges, a 100 % test is conducted. The choice of width of the scan, as defined by British Standard BS 5996 1993, depends on the thickness of the plate. This standard quotes the following examples of scan widths.

- 50 mm for thicknesses ranging from 5 to 33.3 mm,
- 1.5 times the plate thickness for thicknesses of up to 66.6 mm, and
- 100 mm for thicknesses of from 66.6 to 200 mm.

When testing lower grades of steel, the sensitivity of the equipment is adjusted by setting the first back-wall echo from a defect-free area to FSH but, for higher grades of steel, the DAC method (section 8.3.2) is used, with the aid of a 5 mm diameter flat-bottom hole as the reference target. This standard requires the presence of an imperfection to be indicated either by a 50 % reduction in the height of the back-wall echo, disregarding the existence of any flaw echo or, if there a flaw echo having a height equal to or greater than 50 % of that of the reduced back-wall echo. When the DAC method is used, the presence of an imperfection is indicated either by a 50 % reduction in the back-wall echo or by an echo of height equal to or exceeding that of the DAC curve (section 8.3.2). When sizing imperfections using manual scanning the 6 dB drop (half beam) method is used (section 8.2.1). With automatic scanning, the use of a C-scan provides an outline image of any imperfection.

9.2.2 Rolled sections and billets

Rolled sections such as girders and angle iron have characteristic shapes attained with a minimum amount of machining. As a result, a great amount of thought should be taken in the selection of probes and testing procedure especially if there are tapered sections and, in this respect, availability of drawings of the sections are essential to achieve good results. An excellent example of a rolled section, from the point of view of testing, is a railway rail (Fig. 9.1) which would require both compression and angle shear wave probes to obtain a complete test. More than one hundred years of experience of this type of section has given rise to the provision of very well documented evidence of the type and positions of flaws to be expected in service; this provides essential information to the tester (Farley 1978). The rails are tested in the mill when new (Fig. 9.2(a)) and, after installation, at regular intervals with a specially adapted train equipped with either a combination of compression and angle shear wave probes located in specially made sliders (Fig. 9.2(b)) or wheel probes.

In the first rail test car used in the UK in 1971 there were five probes for each rail, one compression wave and four shear wave, two of 35° and the other two of 70°, each located in nylon sliders and irrigated with water when in use. The speed of the train was $40\,km\,h^{-1}$. The information collected by the ten probes was converted from A- to B-scan and images of the B-scan were recorded on 35 mm film. After processing, the film was interpreted by a team of analysts (the author (G.S.) took part in the training of the original participants of this project). After 20 years of service, this test car was replaced by a new

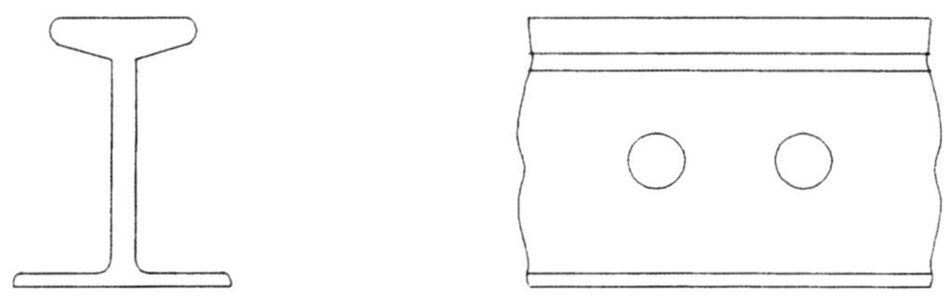

(a) (b)

Fig. 9.1 Rolled section of rail showing: (a) end view, and (b) side view.

(a)

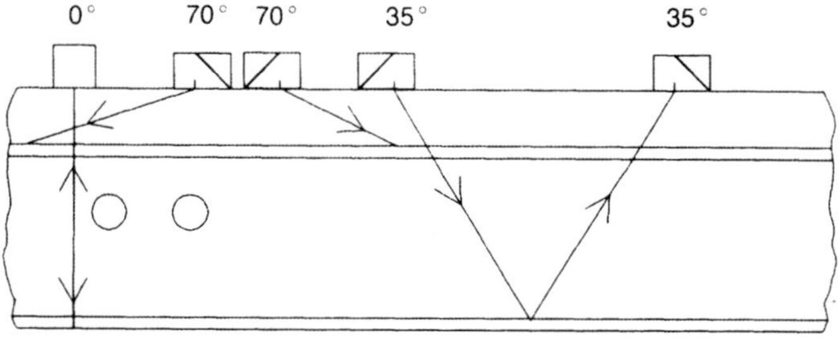

(b)

Fig. 9.2 (a) Automatic testing of rail section after rolling at a mill. (Reproduced by courtesy of Krautkrämer GmbH & Co.) (b) Rail section showing five probes: a compression-wave probe, two 70° shear-wave angle probes and two 35° degree angle shear-wave probes.

one fitted with five wheel probes per rail and having the same angles as the original ones and the speed increased to 64 km h^{-1}. Now the A-scan information is converted to B-scan, as before, but the analysis is performed by an on-board computer.

The use of the ultrasonic rail car has enabled the whole rail network to be thoroughly tested annually, which was not previously feasible. Previously, testing was carried out by men walking along the track to inspect the rails visually. The use of the ultrasonic method has made an immense contribution to rail safety.

Rolled billets are approximately square in section and, although they contain defects similar to those found in rolled plates, they must be tested from all four sides to avoid missing flaws which may occur when testing from only one side (Fig. 9.3). In the rolling mills where billets are produced in long sections, it is usual to have automated means of testing with two probes located on adjacent faces. The techniques used for evaluating flaws, setting sensitivities and choosing probes are similar to those employed for testing plates.

9.3 RODS AND BARS (see also section 7.4)

Rods and bars are produced either by rolling or by drawing through a die. It is assumed here that they differ from one another in that rods have curved surfaces and bars have plane surfaces. The curved surfaces of rods present difficulties in coupling with probes, especially when twin probes are used. In the latter instance, the acoustic barrier between the individual probes should be placed in a direction perpendicular to that of the axis of the rod otherwise there would be no contact between

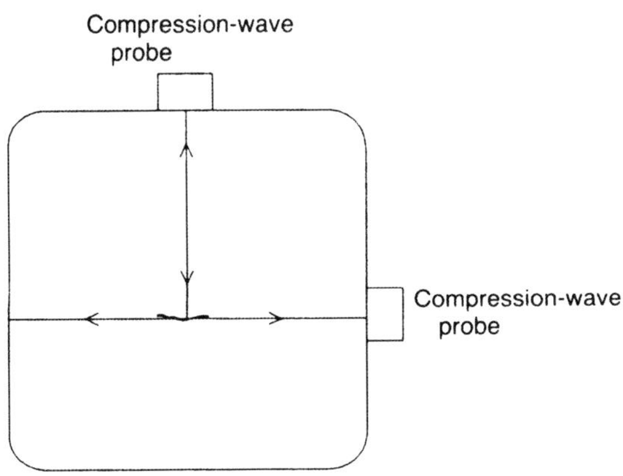

Fig. 9.3 Square billet section showing compression-wave probe scanning from adjacent sides to ensure detection of central faults.

the probes and the rod surface (Fig. 9.4). Because internal defects can lie in any direction, three scans, 120° apart with respect to the axis, should be made in the radial direction (Fig. 9.5) and even if reflections from these defects are not received by the probe, they can be identified from the reduced height of the back-wall echo. For testing bars, scans are required over each of the flat faces.

A common fault found in rods, and also in tubes, is a longitudinal defect called a seam, or fold, resembling a tool mark and resulting from the drawing process during manufacture. Although it may be detected

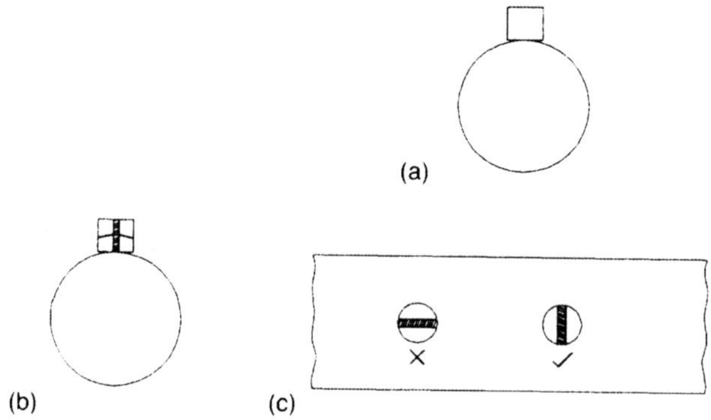

Fig. 9.4 Testing of round bar: (a) with a single-crystal probe, and (b) with a twin-crystal probe, (c) The correct placing of the twin probes, with the acoustic barrier at right angles to the longitudinal axis.

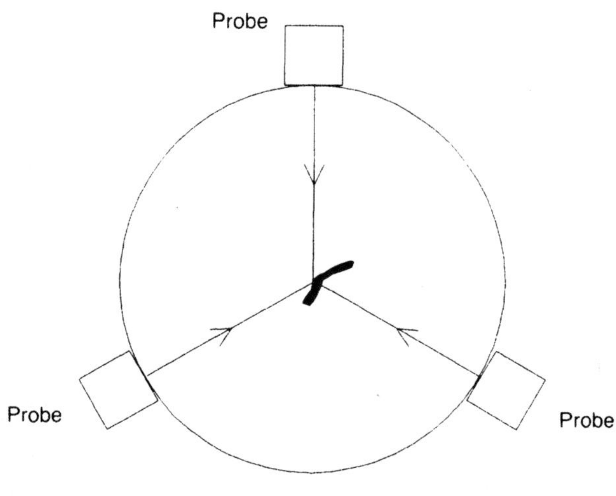

Fig. 9.5 Positions of probes at 120° intervals for testing a round bar.

by eye, it is normal to use an angle probe moved around the curved surface. This lengthy and arduous process is alleviated with the use of automated installations. On reducing diameter, this test becomes more difficult especially when, in the extreme case, the rod is effectively a wire. Wires are difficult to test with piezoelectric ultrasonic probes but they can be more easily tested by using magnetostriction (section 4.5.3) although the use of electrical methods, such as eddy current testing, may be preferred.

9.4 SHAFTS, ROTORS AND ROLLERS

Rotor forgings can range in size up to 10 m or more in length and over 1 m in diameter but shafts are generally smaller: 2 m long and rollers, as used in rolling mills, about 5 m long. Tests on these items can be conducted both longitudinally and across the diameter using compression-wave probes and shear-waves angle probes. On some occasions a single crystal compression-wave probe is mounted on a wedge so as to propagate compression waves at angles of up to 10° to the normal without any mode conversion into shear waves.

Axles of railway stock have long been tested with ultrasound. The simplest test is carried out using a compression-wave probe placed at the end of the axle (Fig. 9.6a) but this does not always produce successful results because of 'blind spots' due to the shape of the shaft. This difficulty has been partly overcome by means of a 10° plastic wedge as described in the previous paragraph (Fig. 9.6b). Otherwise, the use of shear-wave angle probes is required.

For testing axles *in situ*, the probe would generally have to be located

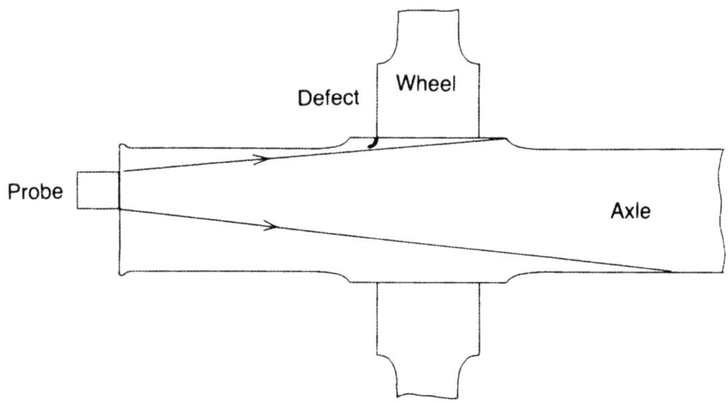

Fig. 9.6a The testing of a locomotive axle a with a shrunk-on wheel by a compression-wave probe positioned where the crack c cannot be detected.

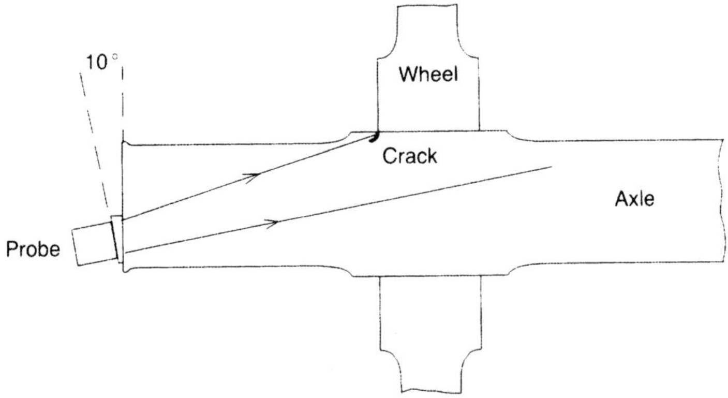

Fig. 9.6b As Fig. 9.6a but with the use of a ten degree compression-wave probe to enable the crack to be detected.

at the ends, which are likely to be the only accessible surfaces. As an aid to testing, a pattern of the echo signals appearing on a flaw-detector screen is prepared for each type of axle in use. The patterns are obtained by a thorough examination at a location, usually a central laboratory, where all of the axle surfaces are accessible. The patterns are analysed and all echoes, real and 'ghost', and direct and mode converted, are evaluated and recorded. At the same time as the procedure for setting up the equipment, a copy of the screen display can thus be made available to the operator who would have received appropriate training in the interpretation of any likely results.

The existence of a shrunk-on wheel does generally hinder axle testing and account should thus be taken of the fact that ultrasound may find its way into the wheel and undesirable echoes caused by compression and mode-converted shear waves may appear. The use of a higher frequency may reduce the effect this phenomenon and thus make it possible to sift out the wanted from the unwanted signals. When the wheels and axles are removed from the rolling stock for overhaul in the railway depot, automatic ultrasonic scanning of the axle seat is possible.

Lifts (elevators) used in public buildings have a life-time of about 20 years and each is expected to be used several million times. Consequently their drive shafts need regular inspection for fatigue cracks. Of necessity, they have to be inspected *in situ* and this is usually done from the ends. Each shaft is of a particular type and the nature and positions of potential failures are known. The resultant information is fed into the memory of the ultrasonic equipment and the maintenance engineer is in a position to select a program to suit the type of shaft and thus conduct the examination. The engineer would have received training in ultrasonic testing relevant to the work and be certified after being examined.

9.5 HEAVY FORGINGS

Heavy steel forgings, which are often more than a metre in thickness, are examined at an early stage in production, but the existence of a rough forged surface and a rather coarse grain structure may require local surface grinding to obtain good probe coupling. Even so, it may be necessary to use a frequency as low as 1 MHz. The grain becomes more refined after heat treatment and the use of frequencies of 2–5 MHz becomes possible, thus allowing for an increased sensitivity of testing. With larger forgings, flow lines are often displayed (Fig. 9.7) and these lines may influence the positions of faults and thus the choice of testing technique because it is essential to ensure that the ultrasonic beam is perpendicular to flow lines. As most of the faults lie along the flow lines, the use of a transverse scan is preferred to a longitudinal scan because the latter is unlikely to indicate any flaws. The ideal time for testing is after the first machining of the outer surface has taken place and before any slots, keyways, grooves and steps are cut.

With a large forging it is sometimes the practice to bore a hole through its centre so as either to remove any central defects or to obtain access. The back-wall echo is often used to establish the sensitivity of testing. When scanning across the diameter of this type of forging, the back-wall, as presented to the ultrasound, is concave in shape when the forging has not been bored but convex in shape when it has been bored. Provided that the diameter is greater than 3.7 times the length of the near field, a concave back wall may be considered to be equivalent to a flat back wall. If the centre hole is used as a back wall, the gain in decibels must be reduced by an amount depending on the ratio of the outer diameter of the forging to the diameter of the bore hole, derived from the nomogram illustrated in Fig. 9.8.

Automatic inspections of this type of forging have been carried out

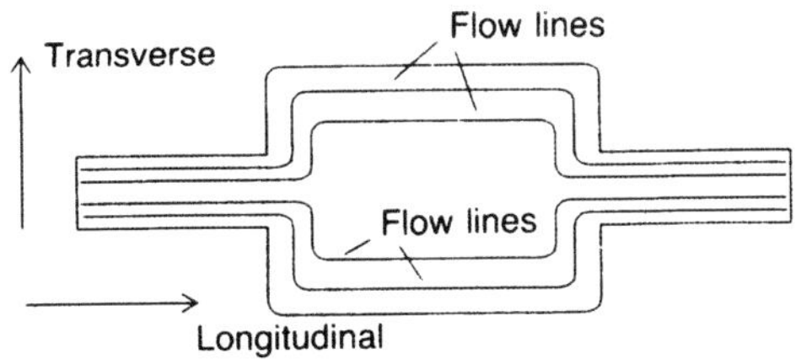

Fig. 9.7 Flow lines in a large forging.

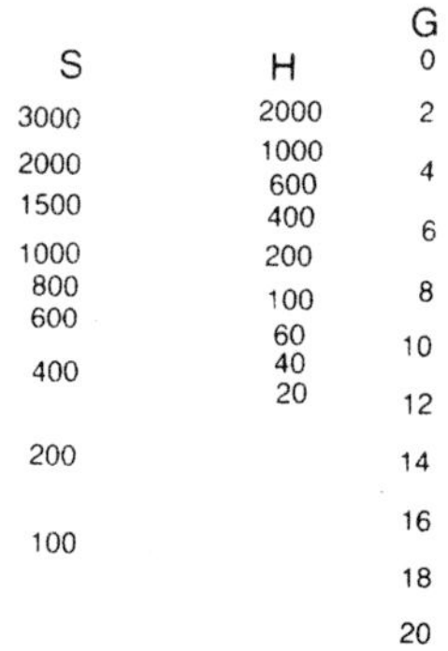

Fig. 9.8 Nomogram for the assessment of the number G of dB to be subtracted from the amplitude of the backwall echo from the inner bore hole when the latter is used to set sensitivity levels. S = Diameter in mm of the forging and H = diameter in mm of the bore hole.

since the early 1960s with the testing performed after initial machining with the component still in the lathe. A probe is fitted to the tool post of the lathe and cutting oil used as a couplant. With the lathe running at its slowest speed, the test can be made with the aid of a monitor which sends a signal to a solenoid in contact with the forging and which operates a marker to facilitate later inspection. Using this method, an inspection originally taking two days is reduced to only 4 hours.

9.6 WELDS

9.6.1 General considerations

Before attempting to test a weld, it is important to obtain information about the preparation and technique relating to it before planning the inspection procedure. For butt welds, it is usual to prepare either single or double vee faces usually at an angle of 60°. The choice between using either single or double vee faces depends on the thickness of the material; material thicknesses of more than 25 mm present problems of access to the welder and thus require double vee faces. Figure 9.9 illustrates a flaw location plotter (section 8.2.2) used to indicate the position of a probe to locate side-wall lack of fusion.

Prior to commencing a weld scan on an object with an angle probe,

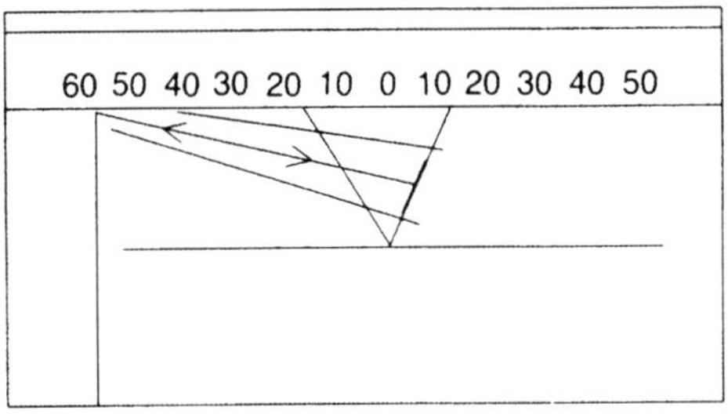

Fig. 9.9 Indications, obtained by a flaw location plotter, during a weld scan.

the parent material of the object close to the welded seam, through which the ultrasonic beam would travel, requires scanning so as to discover any laminar type faults likely to deflect the inclined beams and thus disrupt the testing. The area tested is that bounded by the weld, the edges of the test piece and the line of scanning by the probe. If this surface is in the 'as welded' condition, it may need to be dressed to remove weld spatter and, in any event, to ensure that the transfer losses between the plate and test block (section 8.4) do not exceed 6 dB. During a weld test, the first scan is that of the weld root and is conducted from both sides of the cap at one half-skip distance. Once the centre line of the weld root is established, the half-skip distance is marked, e.g. with a ruler, straight edge or length of tape, so as to provide a guide for the probe as it scans the weld. It is most important here to note the half-skip position on the screen of the instrument so that, during the test, the variable positions of the signals from the root can be compared with the datum line for half skip; this technique is called **point calibration** (Fig. 9.10).

The most fruitful area to search for weld faults is the root where the correctness or otherwise, i.e. either excess or lack of root cracking, can be established. With care, lack of root cracking and root penetration, can be determined by comparing results obtained from both sides of the weld at a given point. If a major defect such as root cracking is found in the root, there may not be any need to test any further.

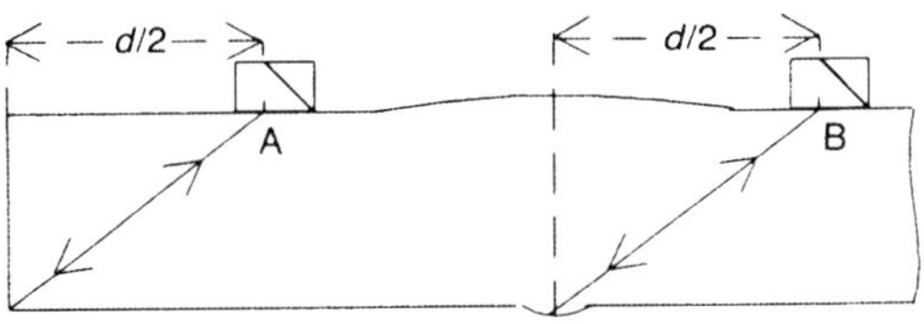

Fig. 9.10 Positions, A and B, of shear-wave angle probes for a point calibration of a butt-welded plate: d/2 = half skip-distance.

Conversely, the finding of a perfectly correct root indicates that the body of the weld may be sound.

9.6.2 Vee welds

Single vee welds, which are most commonly found in pipes and thinner plates, require to be tested at the root, as described above. When welders find difficulties with the root, assistance in the forms of backing bars, backing rings and fusible inserts can be made available (Fig. 9.11). A backing bar is a strip of metal placed in the position of the root so as to allow a sufficient amount of heat to ensure fusion but with no danger of excess penetration. The operator performing a root scan should take the use of this device into consideration and be aware that the half-skip signal should originate from the backing bar (or backing ring if appropriate) and not from a penetrated root bead. The position of the signal from the backing bar, although at a longer distance of half skip from the probe, indicates whether or not the bar is correctly fused. The use of backing devices only interfere with weld examinations when the probe is at the half-skip position.

For **double vee welds**, a characteristic defect is lack of root penetration (Fig. 9.12). It is found at half the weld depth at which the welder has failed to obtain a sufficient amount of root fusion and a planar vertical

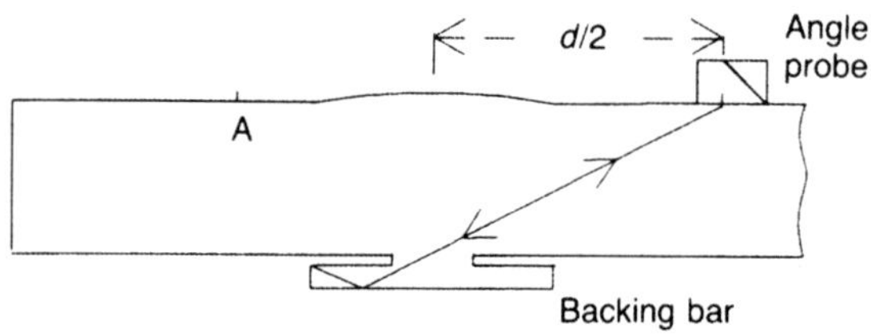

Fig. 9.11 Transmission of ultrasound through a backing bar for weld testing with an angle probe. d/2 = half skip-distance.

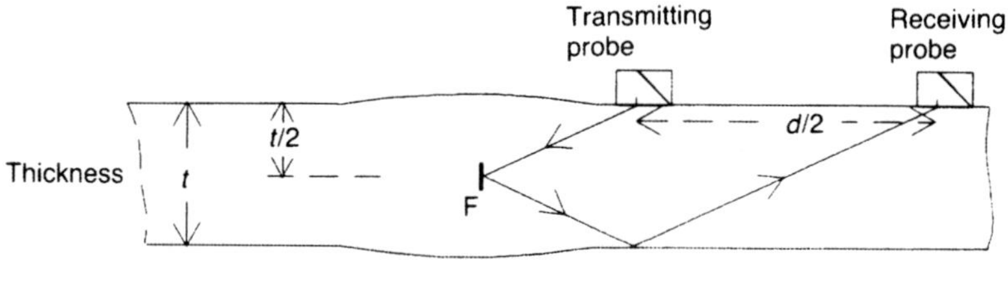

Fig. 9.12 Tandem technique used to detect lack of peretration at F in a doublevee weld. d/2 = half skip-distance.

fault is thus created. Because this defect cannot be detected with a single angle-beam probe, a pair of angle probes must be used, either in tandem, one behind the other (Fig. 9.12), or pitch and catch, on opposites sides of the weld at one skip distance apart (Fig. 8.4).

For satisfactory weld testing, a distance calibration is required; this is usually performed with reference to the centre of the weld cap. The measurements are indicated on a cross-sectional view of the weld so as to aid the exact locating of the reflector (Fig. 9.13). Measurements should always be taken from both sides of the weld. The flaw-location slide (Fig. 8.3) provides an invaluable help to assessing these measurements.

9.6.3 Transverse cracking

Testing for transverse cracking in welds cannot be made by any of the methods mentioned above and special attention is required for this purpose. Provided that the weld cap has been ground flush with the surface of the parent metal, a scan can be made across the centre of the cap with an appropriate probe, e.g. 70°. Otherwise, the scan should be made with the probe inclined at a suitable angle towards the weld, as shown in Fig. 9.14. The movement of the probe can be eased by the preliminary removal of weld sputter. An improvement of scanning is obtained with two 45° probes connected in such a way that one

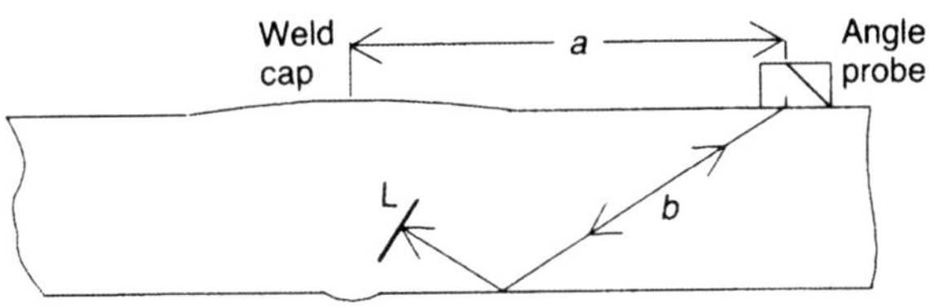

Fig. 9.13 Flaw location with weld testing. L = lack of side-wall fusion, a = distance of probe from weld centre and b = beam path.

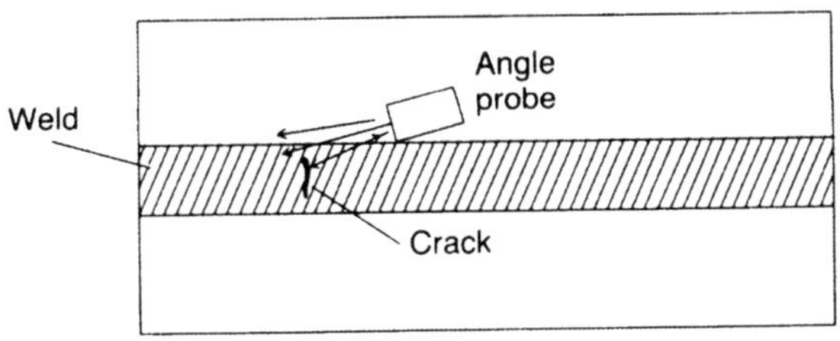

Fig. 9.14 Transverse cracking in a weld detected using an angle probe slightly inclined towards the weld.

transmits and the other receives when a transverse flaw is detected. The two probes are harnessed together and guided in order to obtain a consistent search (Fig. 9.15).

A more elaborate device, as used in pipe mills, is an automatic weld scanner using four probes to form an assembly which is guided along the weld by means of rails (Fig. 9.16). The use of an arrangement of monitor gates and triggers to record gathered information can enable the detection and evaluation of both longitudinal and transverse flaws either by recording the relevant information or operating a paint spray to indicate the position and nature of a detected fault. The paint indications can later be examined manually to provide more detailed evaluations. The latter method is often chosen because of its simplicity and is frequently used for testing machine-made welds such as submerged arc welds.

9.6.4 Tee-butt welds

Welds in which the two components, e.g. plates and ribs, are connected at right angles to one another are called tee-butt welds (Fig. 9.17). The techniques used for testing these welds differ from those for testing plate welds. Tee-butt welds may either be fully or partially penetrated at their roots, depending on the thicknesses of the plates.

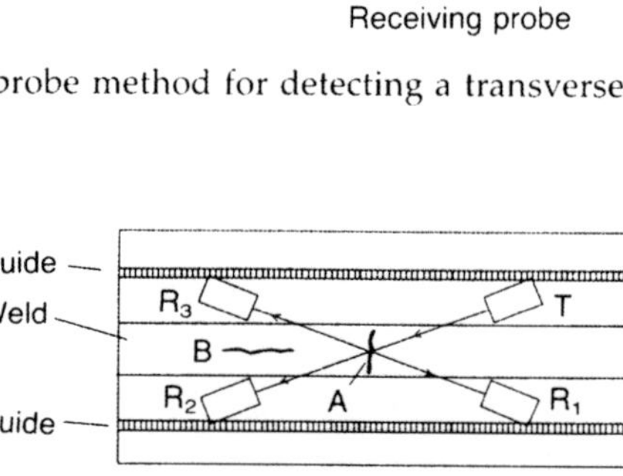

Fig. 9.15 Double-probe method for detecting a transverse crack in a weld.

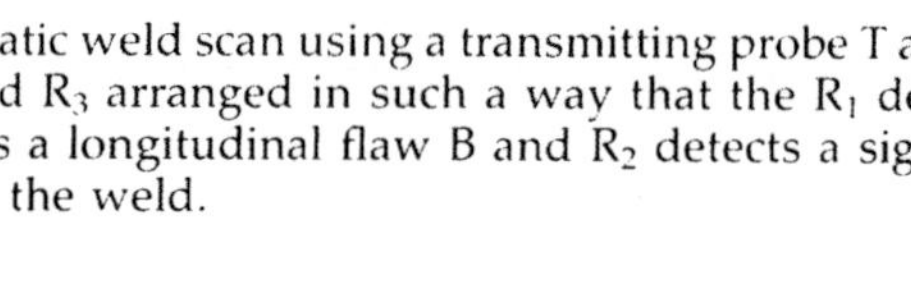

Fig. 9.16 Automatic weld scan using a transmitting probe T and three receiving probes, R_1. R_2 and R_3 arranged in such a way that the R_1 detects a transverse flaw A, R_3 detects a longitudinal flaw B and R_2 detects a signal only when no flaw is present in the weld.

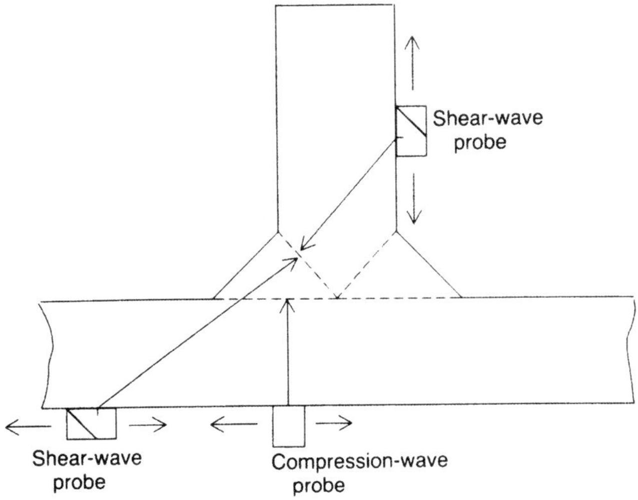

Fig. 9.17 Fully-penetrated tee-butt weld tested with a combination of one compression-wave and two shear-wave probes having different angles.

Partial penetration is often avoided because of potential crack problems at the ends of any unwelded parts. A normal compression-wave probe is often suitable for inspecting these welds and also of any lamellar tearing in the base plate. The latter is caused by the existence of microlaminations in the base plate which may tear in a layer-by-layer manner. If subjected to stress in the direction of the plate thickness, lamellar tearing can occur in tee-butt welds when the base plate is restricted in its movement. The body of the weld is usually tested with either 45° or 60° shear-wave probes to examine for lack of fusion.

9.6.5 Cruciform

In box-section joints, tee-butts are welded back to back to form a joint known as cruciform, which is difficult to test because of the inability to use a normal probe in this instance. Hence only angle probes can be employed for testing this type of joint (Fig. 9.18).

9.6.6 Nozzle welds

A nozzle weld is formed when a pipe is joined to either another pipe but with a larger diameter or a drum. The smaller diameter pipe is known as the branch pipe and the method of joining may be described as either set-in, set-on or set-through, depending on the type of preparation. A set-in joint is one for which the branch pipe is recessed into the main drum and, when the branch pipe sits on the surface of

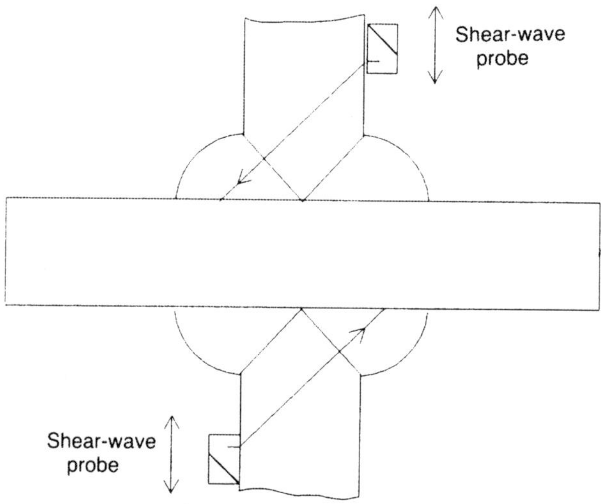

Fig. 9.18 Use of single shear-wave probes to test a cruciform weld.

the main drum, a set-on joint is obtained. A set-through joint is one where the branch pipe passes through a hole in the main drum and welded at both inside and outside the latter (Fig. 9.19). Complications can arise when the branch pipe is set at an oblique angle to either the main pipe or drum (Fig. 9.20).

The nozzle weld is recognized as being the most difficult type of weld to test and only highly skilled personnel are likely to receive appropriate certification. A set-through fully penetrated joint is welded from both sides and the root is completely fused (Fig. 9.20) thus making it easier to test than a joint with an unfused section at the root, i.e. partial penetration. When cross-sectional views are taken around the circumference, a similarity with the tee-butt is seen, particularly when the branch pipe is at right angles to the main drum. A normal

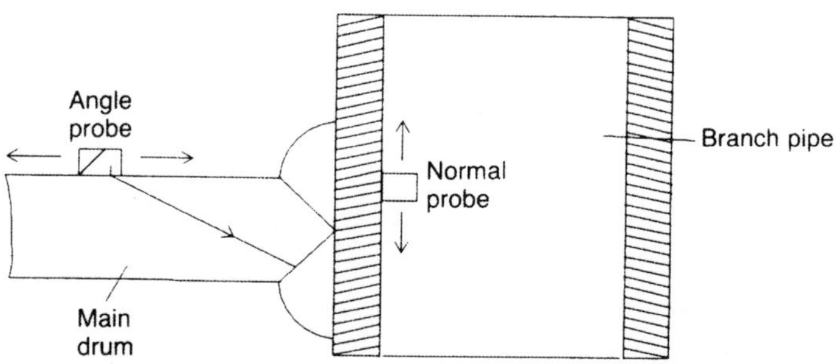

Fig. 9.19 Testing of a nozzle weld.

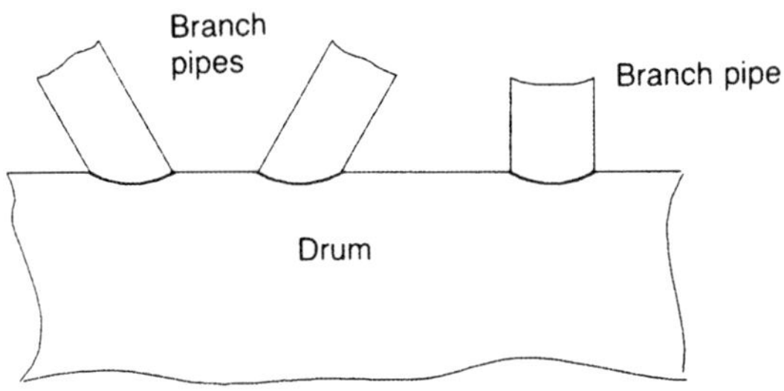

Fig. 9.20 Connections of branch pipe to main drum.

probe can be used to examine one of the fusion faces if access to the inside of the branch pipe is possible. The other fusion faces must be examined by means of an angle probe. It will also be seen that when 'circumnavigating' the branch pipe, the distance from the probe to the weld increases as a result of curvature and it may be necessary to change the probe for one of another angle. The use of a needle or mimic gauge is helpful to obtain sketches of the cross-sectional views required for this kind of test (Fig. 9.21).

When testing welds in austenitic steel with angle probes, it may happen that the ultrasonic attenuation is sufficiently high to reduce the signal to below noise level so that no indications on the screen are observed. This difficulty may be overcome with the use of a compression-wave angle-probe and, because wavelengths of compression waves are about twice those of shear waves, the level of the signal is increased by a factor of about 16, i.e. 24 dB (equation 3.9a). Although angle compression-wave probes have only a limited use, they are obtainable from leading manufacturers.

When flaws are located, their positions should be fully recorded

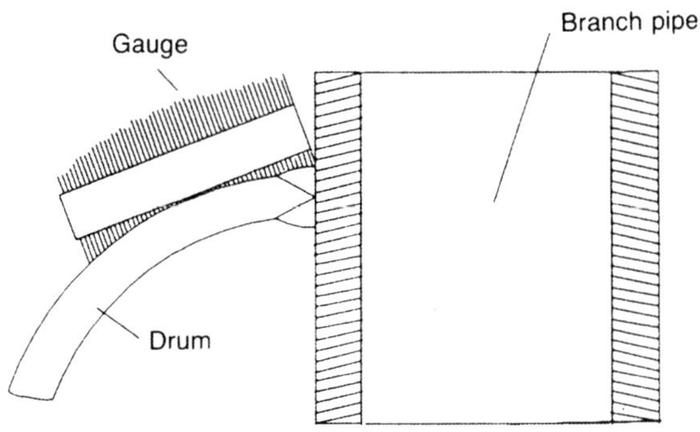

Fig. 9.21 Use of needle, or mimic, gauge to obtain contours of main drum.

in such a way that another qualified person can find them without difficulty. It is common practice to assess the types of flaws from the probe movements (Fig. 9.22). The Welding Institute (IIS/IIW/850/86 1987) lists four of these movements from which the size and properties of a flaw can be determined:

- Lateral, for its length;
- transverse, for its depth;
- rotational, for whether it is either planar or ragged; and
- orbital, for whether it is rounded or not.

It should be remembered that the amplitude and shape of an echo signal play an essential part in the evaluation of the corresponding flaw. Methods for flaw sizing in welds are described in Chapter 8.

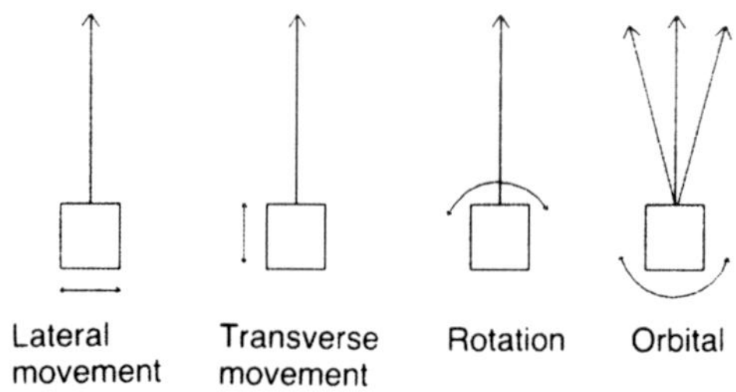

Fig. 9.22 Probe movements.

9.6.7 Spot welds

Spot welds are used to join together metal sheets having thicknesses of 1.5–5 mm by placing one sheet on top of the other and then passing a current through the sheets with the aid of prods located on the opposing surfaces The metal sheets are then fused together at the positions of the prods. The fused area is referred to as either a slug or a nugget and testing for fusion is achieved by passing ultrasound through the joint by means of a small compression-wave probe. The surface suffers deformation because of the mechanical pressure exerted during the welding process and concave depression appears; this makes contact between the probe and surface difficult. When inspecting for lack of bond in the weld, it is important to reduce the width of the beam otherwise signals from the plate interfaces outside the weld are detected and thus misinterpreted as indications of faults. Special types of probe are required for this type of testing and one, which has been used with success, is fitted with a soft membrane on its face and is separated from the surface of the probe by water which acts as a delay path between them (Fig. 9.23).

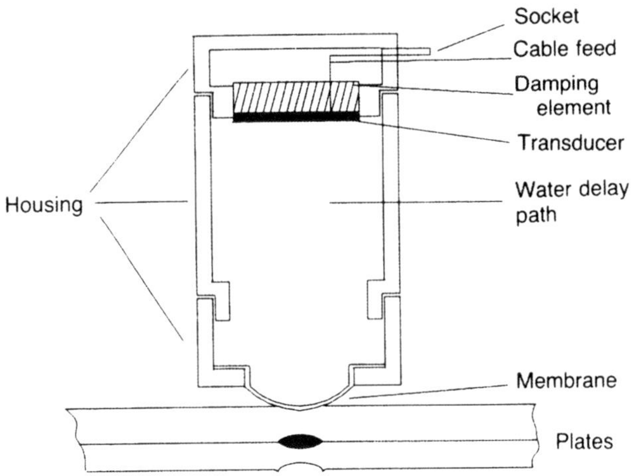

Fig. 9.23 Probe used for testing spot welds. After Frohardt *et al.* (91/2). (Reproduced by courtesy of Krautkrämer GmbH & Co.)

The soft membrane on the probe should fit well into the contours of the spot weld and be kept in a state of tension. If this cannot be achieved, it should be replaced and this would entail refilling the probe with water and ensuring that no bubbles are present. Because a high degree of resolution is essential, the operating frequency should be 15 MHz which is well above normal operating range, 1–10 MHz, used for other forms of ultrasonic testing.

The usual A-scan display for thin plate testing takes the form of closely-packed multiple echoes but, when the probe is located over the spot weld, the echoes become more widely spaced so as to indicate the weld thickness. The echo pattern becomes shorter as a result of increased attenuation caused by the structure of the weld in addition to that caused by scatter. Once the nature of the pattern for a good weld has been established, the pattern characteristic of a poor spot weld will be seen to be distinctly different. Initial calibrations are required with a single plate and then with plates having different types of spot welds. A 95 % rate of successful evaluation can be achieved for this type of testing. Further information may be obtained from Frohardt *et al.* (1990/91).

9.6.8 Procedures for testing welds

A procedure for the ultrasonic testing of welds has been laid down by British Standard BS 3923 Part 1 *Manual Examination of Fusion Welded Butt Joints in Ferritic Steels 1986* and takes the following form:

'11 TEST PROCEDURES TO BS 3923

The BS 3923 requires a test procedure to be prepared for the particular weld or class of welds to be examined:

The following information and instructions shall either be included in the test procedure or be given to the operator through other relevant documentation:

(a) examination level,
(b) type(s) of ferritic parent material and form (i.e. cast, forged, or rolled, etc.),
(c) manufacturing stage at which examination is to be made, and post-weld heat treatment if any,
(d) joint preparation and dimensions,
(e) welding procedure,
(f) surface condition category,
(g) couplant,
(h) type of ultrasonic flaw detector,
(i) details of scans and probes to be employed,
(j) method of sensitivity setting, and scanning and evaluation levels to be applied,
(k) design and use of special test blocks if any,
(l) method to be used for reinforcing positions of imperfections if found,
(m) reporting requirements,
(n) acceptance standard and
(o) operator qualification

Immediately prior to examination, the operator shall be informed of any local deviations from joint preparation drawn in the welding procedure, and shall be given details of any repairs already made.

Preparation of Results

For each weld examined, the operator shall submit a signed report containing the following information:

(a) job identification and material,
(b) identification of detailed test procedure,
(c) stage of manufacture at which examination was made, including state of heat treatment,
(d) identity of operator and certification status,
(e) date of test,
(f) sketch of weld configuration as inspected showing all relevant dimensions, and one or more cross-sectional views,
(g) type of ultrasonic flaw detector used (including serial number),
(h) details of probes used, including make, type and frequency,
(i) calibration block used,

(j) test sensitivity and evaluation/recording levels applied,
(k) method of correction employed for attenuation and transfer loss,
(l) sketch showing extent of scanning, in particular, detailing areas which, due to the geometry or surface condition of the component, could not be fully examined in accordance with the detailed test procedure,
(m) a report on the parent metal examination describing any local heterogeneities (e.g. laminations, surface imperfections, areas of high attenuation) which may adversely affect the weld examination at such locations; if no such heterogeneities are found, a statement to this effect should be indicated in the operator's report,
(n) test results in accordance with the requirements in table 30' (see Table 9.1), 'and including imperfection location in relation to the reference datum points shown in the detailed test procedure,
(o) for flaws which lie outside the acceptance criteria, a plotted illustration of such flaws.

Table 9.1 Extract from BS 3923 PART 1 (1986): table 30 imperfection information required to meet examination levels

Examination level	*Type of imperfection*	*Information to be determined*
1	Point	Maximum echo heigh
	Thread-like	Maximum echo height and length
	Volumetric	Length and maximum cross-sectional dimensions
	Planar	Length and through-thickness dimensions
	Multiple	Enclosing volume and average echo height
	Root profile	Length and depth
2	Point	Maximum echo height
	Thread-like	Maximum echo height and length
	Volumetric	Length and maximum cross-sectional dimensions
	Planar	Length and through-thickness dimensions
	Multiple	Enclosing volume and average echo height
	Root profile	Length and maximum echo height
3	Point	Maximum echo height
	Thread-like	Maximum echo height
	Volumetric	Length and maximum echo height
	Planar	Length and maximum echo height
	Multiple	Enclosing volume and average echo height
	Root profile	Length and maximum echo height
4	Point	By agreement
	Thread-like	By agreement
	Volumetric	By agreement
	Planar	By agreement
	Multiple	By agreement
	Root profile	By agreement

The operator's report shall either include a statement regarding the acceptability or otherwise of each of the recorded imperfections in relation to the acceptance standard or shall be accompanied by such a statement by a supervisor or other responsible person.'

9.7 CASTINGS

Castings are less commonly tested by ultrasonic methods than welds and forgings because their structure is usually such as to provide a high ultrasonic attenuation. A casting is formed by the pouring into a mould of a molten metal which is then allowed to cool during which time it solidifies and a grain structure is formed, the nature of which depends on the shape of the casting. Flaws may arise during this process and they include shrinkage, porosity, the formation of cavities and hot tearing, none of which are formed in the forging stage. These faults give rise to scatter, reflection and attenuation of ultrasonic beams. Reflected beams may not often be returned to the probe because of any random directions of the flaws. Because of the high attenuation coefficients, it is necessary to utilise lower frequency probes to ensure complete penetration of the material being tested but a poor degree of resolution may result. The use of a through transmission method (section 5.4) can detect a faulty area but provides no information as to its depth.

Radiography is often the preferred alternative to ultrasonic testing but the choice of techniques is limited by the thickness and density of the casting. Inherent problems with ultrasonic testing of a casting are that its shape usually includes curves and tapers and there is the possibility of a rough 'as cast' surface, especially with sand castings. Nevertheless, castings are often tested with ultrasound but not necessarily for detecting flaws. For example, 'core shift', which arises when a core placed in a mould to create a particular internal feature moves when the metal is poured into it. One application of this test is the examination of cast motor-engine blocks to ensure that the thickness of the cylinder wall is both correct and uniform. Variations in thickness can result in differential temperature variations around a cylinder when in service, which are potentially dangerous.

The ability to test castings depends on the type of metal and, for a given type of metal, there can be a large number of related alloys. Cast iron specified by British Standard BS 1452 (1990) is a good example of such a metal. There are at least six grades of this metal depending on the content of carbon, which is in the form of flakes giving rise to a great deal of ultrasonic scatter. The lowest grade of this alloy is virtually impossible to test but this should not give any need for concern because it is used for making non-critical items such as door stops. The highest grade, however, which is often used for making pedestal bearings in power stations, is frequently tested at frequencies of 2 MHz and above. Spheroidal graphite iron, in which the carbon takes a globular form, can be tested at frequencies of up to 4 MHz.

British Standard BS 6208 (1990) provides guidance for sensitivity levels and limits of planar and non-planar discontinuities for cast steel, either plain carbon or low alloy, but does not apply to austenitic steel castings. As a general rule, cast steels can be tested with ultrasound at frequencies of 2–4 MHz. Penetrability is improved by normalizing immediately after casting. This is achieved by heating to a temperature of approximately 800°C so as to refine the grain and thus allow one to test a large casting. Castings of austenitic steel are often impenetrable, but forgings of this material can usually be easily tested.

The examination of cast steel rolls, which can contain very large thermal stress cracks, has to be undertaken at a stage before expensive machining is carried out on what may prove to be a defective casting. These items may have to be tested at comparatively low frequencies, e.g. 0.5 MHz, both along and across the roll, but frequencies of 1–2 MHz can be used for short ranges. To enable the evaluation of inclined flaws, it is also useful to have available 20 mm angle shear wave probes of frequencies ranging from 1 to 5 MHz.

Aluminium and magnesium castings can easily be tested because of their low attenuation coefficients and degrees of anisotropy. However, alloys of aluminium containing copper, manganese or zinc are very difficult to test. As with austenitic steel, the testing of copper castings is very difficult in contrast to that of copper forgings which is easy. Testing a ship's propeller made from cast bronze which, incidentally, is also tapered, is very difficult. When a bronze casting is sectioned and etched, the size of the grain would be found to be surprisingly large. The reader should be reminded that changes in structure of castings, as with all metals, give rise to a corresponding change in velocity.

It is hoped that the contents of this section have convinced the reader that a thorough knowledge of the casting industry is an essential requirement for anyone engaged in the ultrasonic testing of castings.

9.8 RIVETS AND BOLTS

In the aerospace industry, rivets are often known as fasteners and, like bolts, can be forged and pressed. Ultrasonic tests are usually performed from their ends provided that the latter are suitably flat. Faults discovered at manufacture include core defects in the raw materials and cracking at the heads due to excess pressure. When in service, fatigue cracks may be found in bolts at the root of the thread and, in rivets, in the middle of the shank and under the head. Signals

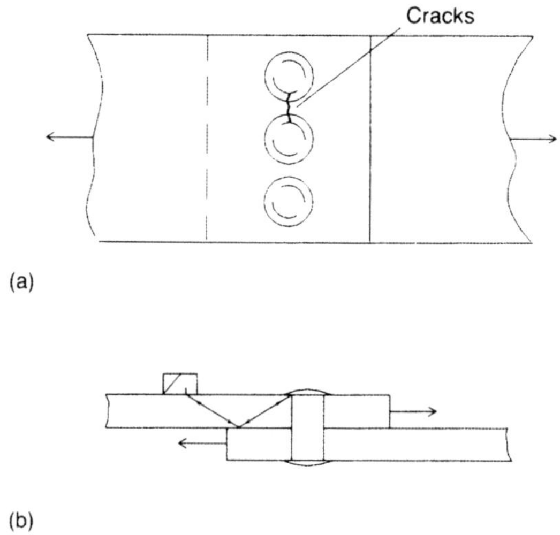

Fig. 9.24 Use of angle probe to test for cracking in a riveted plate (a) horizontal view, (b) vertical view. Arrows denote directions of tensile forces.

from a crack at the root of a thread are not easily distinguished from those from other parts of the thread with the use of a compression-wave probe and an angle probe would be required for this purpose.

When either bolts or rivets are used to fasten plates together, the holes in the plates may develop cracks and an angle probe is necessary to detect them (Fig. 9.24). For this purpose, the hole is scanned at both half and full skip distance and the probe should be swivelled away to determine whether or not a crack is developing from the side of the hole. If the signal from the hole dies away as a result of the swivel, it is clear that there is no crack. Any line of cracking at these holes can be determined from the line of stress in the structure, which is usually easily predicted because the crack takes the least line of resistance on moving from hole to hole, in the same way that perforated paper is torn.

REFERENCES

BS 1452 (1990) *Specification for Flake Graphtic Cast Iron*. British Standards Institute, London.

BS 3923 (1986) *Part 1: Manual Examination of Fusion Welded Butt Joints in Ferritic Steels*. British Standards Institute, London.

BS 5996 (1993) *Methods for Ultrasonic Testing and Specifying Quality Grades of Ferritic Steel Plates*. British Standards Institute, London.

BS 6208 (1990) *Ultrasonic Testing of Ferritic Steel Castings including Quality Levels*. British Standards Institute, London.

Farley, P. (1978) *NDT Internat*, **11**, 287–93.
Frohardt, W., Meincke, F.W. and Voigt, B. (1991/2) *Nondestructive Testing of Spot Welds*. Ref. SD 272/7-0 available from Krautkrämer GmbH, Robert Bosch Strasse 3, D-5030 Hurth 5 (Efferen), Germany.
IIS/IIW/850/86 (1987) *The Evaluation of Ultrasonic Signals*. TWI, Woodhead Publishing, Abington, Cambridge.

CHAPTER 10

The examination of non-metals and adhesive bonds

10.1 INTRODUCTION

Until recently, the ultrasonic testing of non-metals has been largely restricted to concrete. However, with the increase in the uses of new materials, such as fibre-reinforced plastics and the latest improvements in the design of ultrasonic equipment, there has been a significant increase in the use of ultrasound for testing non-metals. This chapter deals mainly with the testing of concrete, rubber, ceramics, plastics and fibre-reinforced plastics. It also includes the testing of adhesive bonds which are, of course, non-metals.

10.2 THE TESTING OF CONCRETE

10.2.1 General considerations

Concrete is a multi-phase material consisting of a coarse aggregate comprising particles of more than 5 mm in diameter, a fine aggregate, sand and cement. The speed of sound in concrete depends on the relative concentration of the constituent particles, the degree of compaction, the moisture content and the nature and amount of defects present. Its coarse granular structure gives rise to a high degree of acoustic scattering and, hence, attenuation. For this reason, testing is usually done at kilohertz frequencies and the pulse through-transmission (or shadow) method (section 5.4) is generally preferred rather than the pulse-echo technique. Complications arise in the testing of reinforced concrete because of the presence of steel rods.

Testing in concrete is usually carried out with the use of compression waves, the speed of which is given by either equation 3.2 or 3.5, depending on the ratio of the lateral dimensions of the test sample to the wavelength. The values of the elastic moduli used here are the dynamic ones, which may be several times greater than the static ones, because of the time lag of the strains behind that of the applied stresses due to the anelasticity of the material. An empirical relationship between the static and dynamic Young moduli, E_{st} and E_d respectively, (BS 1881, 1986) is given by:

$$E_{st} = 1.25E_d - 1900, \tag{10.1}$$

where the values are expressed in megapascals.

The ultrasonic testing of concrete is usually restricted to measurements of the transit times of pulses through the material and, to a lesser extent, determinations of the amplitudes of the received pulses. Applications include measurements of composition, strength, homogeneity, elastic modulus and age, together with the detection of the presence of defects. An account of early work in the field is given by Jones (1962) and information regarding current practice can be obtained from the appropriate standards such as BS 1881 (1986) Parts 203 and 201. Malhorra and Carino (1991) have produced a handbook dealing with the various methods, including ultrasonic ones, for the non-destructive testing of concrete.

10.2.2 The strength of concrete

The strength of concrete is currently defined as the minimum compressive strength required to produce crushing in a given sample. In the UK, the precise terms of the definition are laid down by BS 1881 (1986), Part 203, which requires the test to be made on a 150 mm cube sample, usually 28 days old for ordinary Portland cement or 7 days old for rapid-hardening Portland cement. The value of this strength ranges typically from 20 to 50 MPa (3000 to 7500 lb in^{-2}). In some other countries the test is performed on a cylindrical sample having a height equal to twice its diameter.

The strength of a structure can be measured directly at any time by means of a core test, which involves drilling out a sample from a region where the resulting cavity does not affect the stress distribution in the structure in an adverse manner. However, this procedure is expensive to conduct and cannot be guaranteed to be non-destructive. Furthermore, it is impracticable until the concrete is at least a week old, which precludes its use for monitoring changes in strength in the early stages of setting. Another method of measuring the strength is to conduct a surface-hardness test and to compare the results with samples having known strengths. Although this method is less troublesome and expensive than drilling out cores, it also is unsuitable for testing newly-placed concrete. Furthermore, surface-hardness tests are useless for concrete of more than about 3 years old because of structural changes within the region of the surface.

Ultrasonic testing can be carried out after a period of 6–9 h after the concrete has been placed into position and, because the bulk of the material is being examined and not just the surface, no time limitation for testing is necessary. Here the speed of compression waves is compared with those in samples having known strengths. Tests of this nature, using a pulse technique, have been summarized by Whitehouse (1951) (Table 10.1). Vold and Hope (1978) have compared the results of later work by a number of investigators (Ruppert and Hess, Elvery, and Davis and Robertson) who obtained curves relating the compression wave speeds to crushing cube strength (Fig. 10.1). The curves, although

Table 10.1 Relationships between the group velocity of compression wave pulses and concrete strength (after Whitehurst, 1951)

Group Velocity ($m\,s^{-1}$)	*Strength*
above 4500	Excellent
3600–4500	Generally good
3000–3600	Questionable
2100–3000	Generally poor
below 2100	Very poor

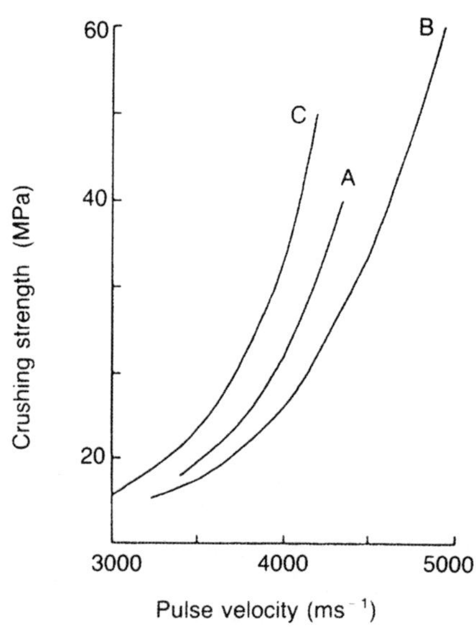

Fig. 10.1 Relationships between pulse velocity and crushing strength for concrete: A = Ruppert and Hesse; B = Elvery; C = Davis and Robertson. (after Vold and Hope 1978)

not coincident, follow a similar trend in which there is an approximately exponential increase in strength with acoustic speed.

Discrepancies between them are attributed to differences in type and quality of cement, type and gradation of aggregate, mixture ratio, conditions of curing and age. Elvery's results, for example, show that the strength, σ, varies with the speed, c, as:

$$\sigma = A \exp Bc, \tag{10.2}$$

where A and B are constants having values dependent on the factors mentioned above. The exponential nature of this relationship indicates the difficulty in assessing the higher values of strength resulting from a small change in c which can be brought about equally by small variations in temperature and moisture content and a comparatively high change in strength.

10.2.3 The effects of age

The strength of concrete increases with age and it is thus important to predict its value at any given stage of a construction process. Elvery (1976) investigated the variation of both the speed of ultrasound and the equivalent cube strength with age and found that the speed increases rapidly with early ageing with the rate of increase diminishing with advancing age (Fig. 10.2). On the other hand, the cube strength remains constant at first and then rises rapidly while the rate of increase in the acoustic speed slows down.

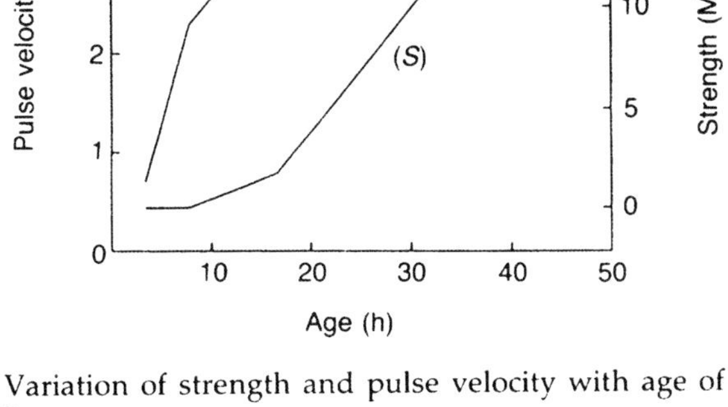

Fig. 10.2 Variation of strength and pulse velocity with age of concrete (after Elvery 1976)

10.2.4 The effect of steel-bar reinforcement

When measuring the ultrasonic speed in concrete reinforced by steel bars, a reduction of the time of transit from that observed for non-reinforced concrete occurs because some of the sound energy travels through the bars, in which the speed is greater than that in concrete. The amount of this reduction depends on the angle of orientation of the axes of the bars with respect to the direction of ultrasonic propagation; it is a minimum for propagation at right angles to the axes and a maximum when the propagation is in the axial direction. In the former instance the reduction is generally small but, if the lengths of the bars are very much less than their diameters, a considerable increase in the speed of ultrasound is observed (Fig. 10.3). If c and c_s are the respective speeds in concrete and steel, l the distance between the transducers, and a the perpendicular distance between the axis of the ultrasonic beam and that of the nearest steel bar, it is shown in Appendix F (equation F.5) that the transit time t is given by:

$$t = 2a(c_s^2 - c^2)^{1/2}/(cc_s) + l/c_s. \tag{10.3}$$

The effect of the steel bar disappears when the ratio a/l is sufficiently high for the transit time along the direct path TR to become less then along the path TPQR. Here $t < l/c$ and it can be inferred from equation F.6 that:

$$a/l > [(c_s - c)/4(c_s + c)]^{1/2}. \tag{10.4}$$

Thus, for example, if the speed of sound in the concrete is equal to $3500\,\mathrm{m\,s^{-1}}$ and that in the steel bar is $5500\,\mathrm{m\,s^{-1}}$, the critical value for a/l is 0.24.

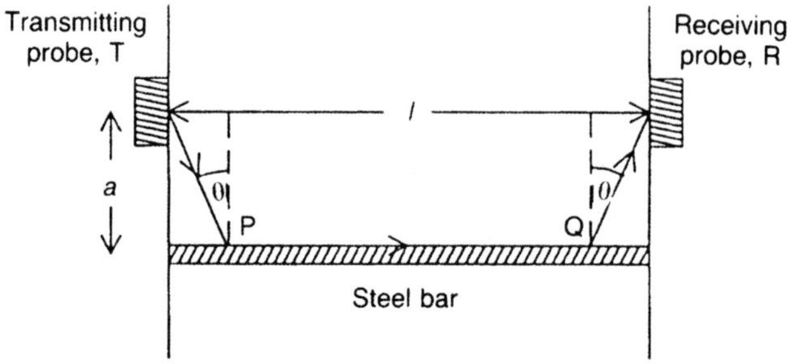

Fig. 10.3 Ultrasonic propagation in reinforced concrete in the direction parallel to the steel bars: a = perpendicular distance from the axis of the direct ultrasonic beam from the nearest steel bar b; l = distance between the probes. (after BS 1881, 1986, Part 203)

10.2.5 The effect of the mix ratio

It is easily seen that the speed of sound in concrete is dependent directly on the relative concentrations of its components, for which the physical properties remain the same regardless of the degree of mixture.

$$1/c = A/c_A + B/c_B + [1 - (A + B)]/c_p. \tag{10.5}$$

Here, A and B, respectively, are the volume fractions of coarse and fine aggregate in a concrete sample and c_A, c_B and c_p are the corresponding speeds of ultrasound in coarse aggregate, fine aggregate and cement, However, in practice, other factors such as the degree of compactability, the water–cement ratio and the curing temperature should be considered (Jones 1962).

10.2.6 Ultrasonic measurements in concrete

The presence of coarse aggregate, often exceeding 10 mm in dimensions, requires that ultrasonic testing in concrete be conducted at relatively low frequencies in order to avoid excessive attenuation caused by scattering (section 3.3). In practice, the chosen frequencies lie between 25 and 250 kHz, corresponding to wavelengths ranging from 200 mm down to 10 mm, depending on the nature of the sample being tested. At the lower frequency limit, it is possible to test concrete structures of up to 13 m in thickness. The measurements made are mostly involved with the speed of sound either to evaluate the dynamic elastic modulus, to gauge thickness or to locate flaws. Because of the predominance of scattering, it is difficult to relate the measured attenuation to any of the inherent properties of the concrete and, consequently, attenuation measurements are usually confined to detection with the pulse through-transmission (shadow) method.

For small test samples, the resonance method (section 5.7.2) can be used for measuring the Young modulus and also the damping capacity, from which attenuation coefficients can be evaluated (Jones 1962). For an accurate determination of the Young modulus, it is necessary for the lateral dimension of the sample to be small compared with the wavelength (section 3.2). Here the sample is prepared in the form of a rectangular block not exceeding 500 mm in length and of cross-sectional area of the order of 100 mm square and clamped at its mid-point (Fig. 10.4). Typically, the sample is energized at one end by means of an electrodynamic oscillator (section 4.5.1), with a lower power similar device at the other end to act as a receiver. The frequency is varied until resonance occurs in the fundamental mode, at which the maximum output of the receiver is observed, so that the length of the sample is equal to one half-wavelength. The 'Q' factor, if required, can be determined by observing the frequency response curve. An

Fig. 10.4 Electrodynamic test bench with contact type vibration generator (left) and receiver (right) arranged for longitudinal resonance tests of a 50 mm × 10 mm × 10 mm concrete prism. (Reproduced by permission of CNS Electronics Ltd.)

alternative method is the through-transmission (or shadow) pulse technique described below.

The **through-transmission pulse method** (section 5.4) is preferred to the pulse-echo technique because of the high attenuation caused by scattering and also the beam spread resulting from the comparatively short near fields characteristic of low frequencies (section 2.7). The received signal may be indicated in the conventional manner as a peak appearing on the screen of a cathode-ray oscilloscope. Alternatively, when only the measurement of the transit time is of interest, a direct-reading digital display can be used. Certain requirements have been laid down by the British Standards Institution (BS 1881, 1986, Part 203) with a view to achieving an accuracy of ±1 % in measuring the transit time of the pulses. The transmitting transducer should provide a rise time of less than one quarter of its fundamental natural time period so as to give a sharp pulse rise (section 2.9) and the receiver should be able to detect the arrival of the earliest part of the pulse even after the amplitude has been highly attenuated. Using a cathode-ray oscilloscope display, the point of intersection of the tangent of the leading edge of the pulse with the horizontal base-line indicates the end of the transit time (section 5.3.2) and, with a digital display, the read-out is simply triggered by the leading edge of the pulse. To allow for time delays in the probes and electronic circuitry, the equipment should be zeroed either by initially bringing the transmitting and receiving probes in contact with one another or by locating them at the opposite ends of a reference bar, having an accurately known length and ultrasonic velocity, and then making the necessary correction.

Accuracy of transit-time measurement can be verified by means of two reference spacers to provide delays of 25 and 100 μs, respectively, (both to within ±2 μs), the shorter spacer being used to check zero time. The measurement obtained should not differ from the indicated values of the reference spacers by more than 0.5 %. The probes, which are usually piezoelectric, should be properly coupled to the concrete surface either with grease, jelly, liquid soap or a suitably prepared paste. Rough and uneven surfaces should be smoothed down. Usually, the pulse repetition frequency is low, e.g. from 10 to 100 Hz, because of the longer ranges as compared with those of tests made with the more conventional ultrasonic equipment, and a typical frequency is 54 kHz, but transducer frequencies range from 25 kHz to 1 MHz, depending on the range and the nature of the concrete. Because much of the work is carried out in the field, the instruments are usually battery operated.

Attention should be drawn to the use of the PUNDIT borehole system (CNS Electronics Ltd, London) for testing concrete piles of circular section, in which a pair of 50 mm diameter tubes are cast in positions close to the opposite ends of a diameter (Fig. 10.5). Pulsed signals are generated by a slightly modified PUNDIT portable ultrasonic flaw detector providing digital indications and designed to operate at frequencies in the kilohertz range. Waterproof transmitting and receiving probes, located in their respective tubes and connected separately by heavy-duty waterproof cable to the instrument, can be raised or lowered simultaneously so as to remain at the same level as one another; acoustic contact is maintained by filling the tubes with water. A digital read-out of the pulse transit time is provided.

Where possible, the probes should be located directly opposite each other on either side of the test sample: the **direct transmission method** (BS 1881, 1986, Part 203), for which the highest sensitivity of measurement is obtained (Fig. 10.6(a)). When access to the opposite surface is not possible, the use of either the **semi-direct transmission method** (Fig. 10.6(b)) or the **indirect or surface transmission method** (Fig. 10.6(c)) is necessary. These latter methods, which would be out of the question for use at megahertz frequencies, are feasible for the lower kilohertz range because of the wide beam spread. Figure 2.12b(i) (section 2.7) illustrates the directivity curve for $ka = \pi/2$, for which the probe radius is equal to a quarter-wavelength. With a 50 mm diameter probe

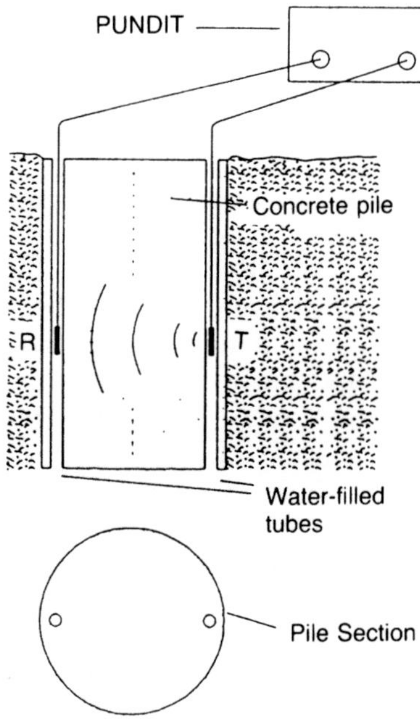

Fig. 10.5 PUNDIT borehole system: T = transmitting probe and R = receiving probe. (Reproduced by permission of CNS Electronics Ltd.)

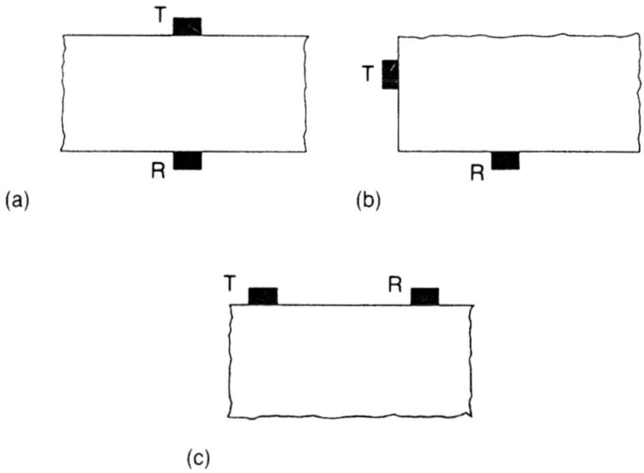

Fig. 10.6 Probe positions for the ultrasonic testing of concrete, T = transmitting probe and R = receiving probe (after BS 1881, 1986, Part 203): (a) direct transmission method; (b) semi-direct transmission method; (c) surface transmission method.

the wavelength is thus equal to 100 mm corresponding to a frequency of 45 kHz assuming a speed of ultrasound of 4500 m s^{-1} (the length of the near field, a^2/λ, is seen to be only 25 mm).

With the indirect method, the bulk compression waves are only picked up because of the large amount of beam-spread characteristic of low frequencies and the received amplitude is thus very small, about 2 or 3 % of that attainable by the direct method. Surface waves may also be also propagated but, because they travel at about half the speed of that of the compression waves, the latter arrive at the receiver first. However, surface waves have a considerably higher amplitude than compression waves because the spread is restricted to only two dimensions and, for this reason, it is essential to verify that the sensitivity of the receiver is high enough to detect compression waves. It should be noted that the indirect method evaluates ultrasonic velocities only for that part of the material close to the surface and its value may be very different for regions remote from the surface. This method, however, is perfectly adequate for surface crack detection for which the presence of a defect interrupts the flow of the waves. However, one must take into account the dimensions of the probe surfaces because of the uncertainty of the value of the ultrasonic path length between them. This can be done by measuring transit times of the pulses for different distances of probe separation and plotting these quantities on a graph. The best straight line is then drawn through the plotted points and the gradient of this line gives the relationship between probe separation and transit time, i.e. the speed of ultrasound.

10.2.7 Flaw detection in concrete

With the through-transmission method, the presence of internal defects in a sample of concrete gives rise to a decrease in amplitude of the received signal. The lateral size of an extended defect can be estimated by scanning with the two probes in tandem, but this method does not indicate the distance of the defect from the surface. It is, however, possible to use the pulse-echo method, with the two probes positioned side-by-side to locate a large crack parallel, or nearly parallel, to the surface provided that the depth allows a measurable echo to be received. Because of the comparatively long wavelengths, it may not be possible to detect fine cracks; referring to equation 2.46, $R_1 = R_2$ and, because l is very small compared with wavelength (typically by a factor of 0.001 or less) there is virtually 100 % transmission across a fine crack even at a frequency of as high as 50 kHz.

Surface cracks, however, are comparatively easy to evaluate with the procedure laid down by BS 1881 (1986), Part 203. The transducers are placed on the surface at equal distances, a, on either side of the crack. In the absence of any defects, surface waves pass from the transmitting to the receiving probe. If, however, a surface breaking crack appears between the probes, it provides a partial barrier to the surface waves, some of which may follow the profile of the crack. However, bulk compression waves arriving at the tip of the crack suffer diffraction and some of them are picked up by the receiving probe (Fig. 10.7(a)). The transit times t_1 and t_2 are measured for two values of a, a_1 and a_2, and it can be shown from simple geometry that the crack-depth, d, is given by:

$$d = [(a_2^2 t_1^2 - a_1^2 t_2^2)/(t_2^2 - t_1^2)]^{1/2}. \tag{10.6}$$

If, for example, we put $a_1 = 150\,\text{mm}$ and $a_2 = 300\,\text{mm}$, we have:

$$d = 150[(4t_1^2 - t_2^2)/(t_2^2 - t_1^2)]^{1/2}. \tag{10.6a}$$

It should be noted that the speed of surface waves is about one half that of bulk compression waves and should the surface waves pass over the crack, the diffracted compression waves would be received first.

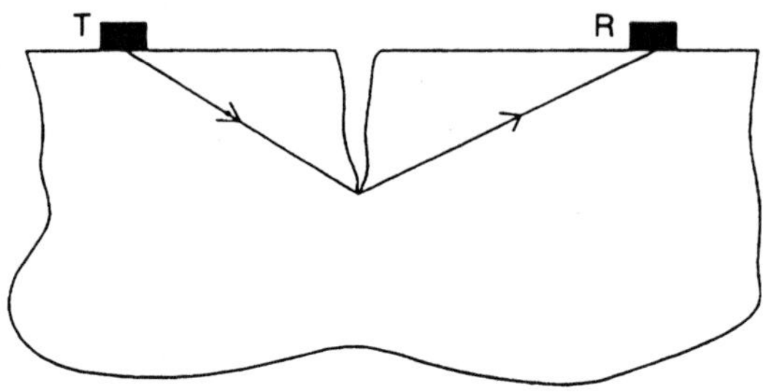

(a)

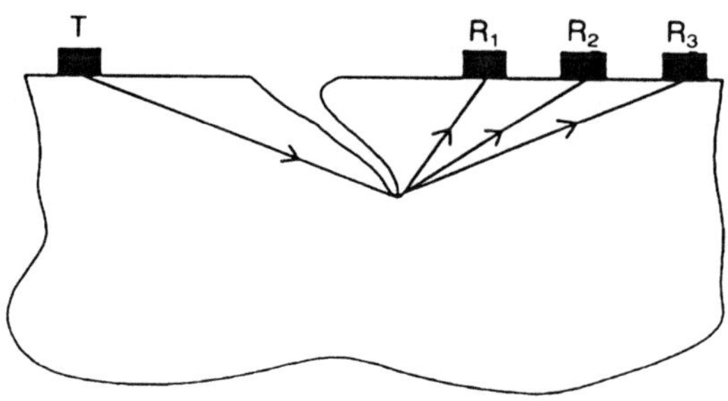

(b)

Fig. 10.7 Detection of a crack in a concrete sample: (a) perpendicular to; and (b) inclined to the surface, T = transmitting probe, R = receiving probe, R_1, R_2 and R_3 = alternative positions of the receiving probe (after British Standard BS 1881, 1986, Part 203)

Equation 10.6 was derived by assuming that the crack is perpendicular to the surface. This can be checked by moving each probe in turn away from the crack. A decrease in transit indicates that the crack inclines towards the direction in which the probe is moved (Fig. 10.7(b)).

When a concrete surface is exposed to the weather, especially during frost, or to fire, it deteriorates in quality. The affected region is found to have a lower ultrasonic speed, c', than the unaffected region which still retains its speed, c. The depth of affected region can be measured in the following manner (BS 1881 (1986) Part 203). Ultrasound is transmitted from the transmitting probe (Fig. 10.8) in all directions, some of which is received directly by the probe R on the surface. The remainder strikes the boundary between the deteriorated and the undeteriorated concrete at a point A, with an angle of incidence equal to or greater than the first critical angle, θ_c, where $\theta_c = \sin^{-1}(c'/c)$ (section 2.5.2), and then travels along the boundary. Owing to the beam being divergent, some of the wave energy is propagated in the undeteriorated region at a speed c, where it suffers progressive refraction at the various points B, and arrives at the receiver in the corresponding position R. When R is close to the probe, at R_1, the earliest received compression-wave pulses pass only through the region just below the surface. However, because c is always greater than c', there must be a position R_2 where the transit time through the surface region is equal to that for the path TAB_2R_2 indicated on the diagram. For any position R_3 of the receiver beyond R_2, the ultrasound travelling along the path TAB_3R_3 arrives earlier than that moving along the direct path TR_3. Thus, when the receiver is moved progressively away from the transmitter in equal increments, the transit time increases by correspondingly equal amounts but, beyond R_2, the time intervals abruptly reduce to a lower but constant value. If x is the distance TR_2 between the receiver and transmitter when the transition occurs, it can easily be seen, bearing in mind that

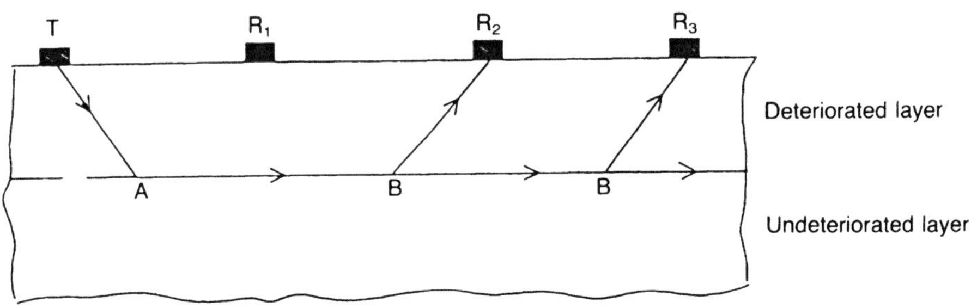

Fig. 10.8 Determination of the thickness of the deteriorated layer of concrete: T = transmitting probe and R_1, R_2 and R_3 = positions of receiving probe.

$$\sin\theta_c = c'/c.\ \cos\theta_c = (1 - \sin^2\theta_c)^{1/2} \text{ and } \tan\theta_c = \sin\theta_c/\cos\theta_c,$$

the thickness, d, of the deteriorated layer, D, is given by:

$$d = (x/2)[(c - c')/(c + c')]^{1/2}. \tag{10.7}$$

This method is effective only when the deteriorated region is extensive, of fairly uniform thickness and of a quality sufficiently reduced for c and c' to be appreciably different from one another.

10.3 THE TESTING OF WOOD AND TIMBER

10.3.1 General considerations

Wood is a highly anisotropic and generally inhomogeneous material. It is cellular in form and has a granular structure. The strength of a piece of timber is high in the direction of the grain but low perpendicular to it. The presence of knots provides weak points. The strength of a timber structure increases with dryness.

Ultrasonic tests can be made on both standing and felled trees and also on cut timber, composite timbers (such as plywood) and wooden structures. They include investigations of mechanical properties, glue bonds and the presence of defects. Speeds of ultrasound vary from 1500 to 600 m s^{-1}, for compression waves, and from 1200 to 2500 m s^{-1} for shear waves, depending on the type of wood, moisture content and the direction of propagation with respect to the grain. Elvery and Nwokoye (1970) have established a high degree of correlation (99.4 %) between the ultimate compressive strength and compression-wave characteristic impedance for a number of soft and hard woods, and Nwokeye (1971) has found a linear relationship, with a correlation coefficient of 98 % between the ultimate tensile strength along the grain direction and the shear-wave characteristic impedance. Elvery and Nwokoye (1970) have also found that for variations of grain angle, the compression wave speed and the ultimate compressive strength follow similar trends. For example, in Douglas fir, the speed reduces from about 6000 m s^{-1} for an angle of zero to about 2000 m s^{-1} for an angle of 90°.

The granular structure of wood gives rise to a considerable amount of ultrasonic scattering, which is not uniform with direction because of the anisotropy of the material, and it follows that the attenuation coefficient also varies with direction with a maximum value across the grain and a minimum along the grain. The upper usable frequency limit is around 1 MHz and equipment originally designed for testing metals can often be employed for testing wood. On the other hand, much work has been done using the pulse through-transmission equipment intended for testing concrete at frequencies of about 40 kHz.

Bucur (1983) has conducted measurements on borings from living trees to determine values of elastic constants with the use of this latter method.

The material used for coupling the probe to the test sample should not have any properties which adversely affect wood and suitable examples include petroleum jelly, glycerine and aluminium soap gel. The latter is especially suitable in that it neither is absorbed by nor stains wood. There is, of course, usually no difficulty in attainment of a smooth surface for timber and the use of immersion techniques is feasible. Shear wave propagation can be achieved with the use of conventional shear wave probes (section 4.4).

As with other materials, the mechanical properties of wood can be investigated by measuring the speed of sound, which is best done with a pulse method. The pulse through-transmission technique, using a frequency of about 50 kHz, is suitable for large samples in the same way as with concrete (section 10.2.6). At this frequency, the waves are propagated in all directions and the use of the indirect transmission method (Fig. 10.6 (c)) is possible; this arrangement is highly suitable for testing planks. For smaller samples, the conventional ultrasonic flaw detector can be employed with the use of frequencies of between 0.25 and 0.5 MHz, preferably with pulse through-transmission. Flaw detection is usually conducted in the higher frequency range, with either the echo or through-transmission method, depending on the acoustic path length, using either A- or C-scans (section 5.6).

10.3.2 Flaw detection in wood and timber

Defects likely to occur in wood and timber include the following.

- **Knot:** the centre of the growth rings in the main trunk and branches of the tree, characteristically of different structure, grain direction and resin content from the rest of the wood.
- **Pinholes:** long bore holes of small diameter and caused by the action of larvae, such as woodworm.
- **Pitch:** accumulation of resin, frequently laminar in structure and usually orientated in the direction of the grain.
- **Rot:** cavities formed by insects or of leakage of moisture and subsequently enlarged by decomposition.
- **Shake:** a split along the grain which becomes filled and may contain some liquid.

In addition, lack of bonding may occur in plywood and other composite wood structures.

Apart from knot and pitch, the defects listed above occur usually in the form of voids, which can be detected by the through-transmission method. However, the location and characteristics of defects can only be achieved in a satisfactory manner with the pulse-echo method operated at the highest possible frequency.

Makow (1969) has investigated the various kinds of defects in hemlock using the pulse-echo method. Suitably prepared samples 40 mm thick were immersed in water and tested with 1 MHz focused transducers. Rot containing air and located 5 mm below the surface gave rise to almost 100% reflection at its upper surface and no back echo could be observed. The defect was characterized by an irregular structure in the appearance of the received pulse on the oscilloscope screen, as a result of the softness of the surface. Rot filled with liquid, however, has been found difficult to detect.

Shake, as a result of its well-defined structure, gives rise to clear echoes of amplitude varying with orientation. Because of the relatively small change of characteristic impedance at its interface with the main structure, pitch does not produce any echoes. However, pitch is not granular in structure and it causes a smaller degree of attenuation than caused by defect-free wood. The presence of pitch is thus characterized by an increase in the amplitude of the back-wall echo. Echoes are observed from knots but, because they attenuate sound waves less than the wood itself, back-wall echoes are again seen.

Plywood consists of thin layers of wood bonded together with glue, with the grains of each layer orientated a right-angles to its neighbouring layer. The most common defects in this kind of structure are lack of bonding and weakness of bonding and tests for these defects can be performed using techniques suitable for fibre-reinforced plastics (section 10.5.3), usually employing C-scanning.

10.4 THE TESTING OF PLASTICS AND RUBBER

10.4.1 General considerations

Plastic and rubber materials are polymeric in structure and are characterized by their anelastic properties, i.e. strains lag behind the applied stresses (section 10.2.1). This property, which is also characteristic of concrete, results from a phenomenon known as creep and, as with concrete, the dynamic and not the static modulus is the one which characterizes the speed of sound (section 3.2).

With variations of temperature and frequency over the ranges, polymers undergo a number of transitions (Read and Dean (1978)); from the glassy state at the lower end of the scale, through the glass-rubber transition region and the 'rubbery plateau', to the flow region where the lattice structure breaks down. These stages give rise to progressive decreases in elastic modulus, coupled with increases in attenuation.

With the exception of fibre-reinforced plastics, which are treated separately (section 10.5), and rubber used for automobile and aircraft tyres, polymer materials rarely form part of critical structures and are therefore not often subjected to high calibre flaw detection. If, however, ultrasonic flaw detection is performed, account should be taken of the relatively high attenuation coefficients when considering the testing parameters. Mention should be made that Perspex samples containing artificial defects are often used to investigate flaw-detection techniques with the use of ultrasonic visualization (section 4.5.5). In general, however, ultrasonic testing of polymers is confined to measuring speeds and attenuations in small samples to determine their physical characteristics, including those outlined above, and also the presence of impurities, porosity and the states of cure of resin-like substances.

10.4.2 Measurements in polymers

Methods suitable for measuring ultrasonic speed and attenuation in polymers are largely the same as those used for metals (Chapter 5). The pulse through-transmission technique and the goniometer are especially suitable, although the use of the pulse-echo method is possible for samples of suitable thickness or having comparatively low attenuation coefficients.

Read and Dean (1978) have described a method specially developed for measuring ultrasonic speed and attenuation in polymers. The sample, in the form of a rectangular block, is mounted on a turntable which can be rotated in steps of 0.01° and located between a pair of heavily-damped compression wave probes, one transmitting and the other receiving. The complete arrangement (Fig. 10.9) is immersed in a

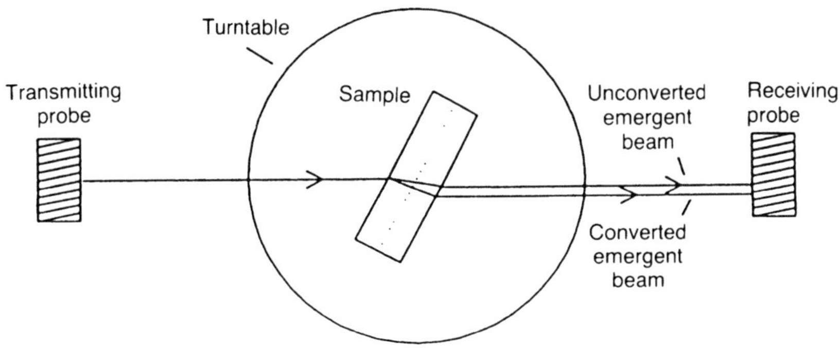

Fig. 10.9 Arrangement for ultrasonic measurements in polymers (Reproduced with permission from Read and Dean 1978)

water tank. Short pulses generated by the transmitter are picked up by the receiver after traversing the sample but, for materials having a sufficiently low attenuation coefficient, the receiver is replaced by a metal reflector parallel to the transmitter, so as to enable the use of the pulse-echo method. When the sample is rotated through a given angle, mode conversion takes place (section 2.5.2) and both compression and shear waves pass through it. Consequently, two sets of compression waves emerge, one which has undergone mode conversion twice (compression waves to shear waves and back to compression waves) and the other which has not; there is thus a time delay between their respective pulses. With the two-probe method, the angles of rotation should be small enough to allow both the lateral displacements of the emergent pulses due to refraction in the material to be picked up by the receiver. This restriction is unnecessary when a reflector is used in place of the receiving probe. The probes are connected to a conventional pulse-echo ultrasonic set (Fig. 5.1) and the peaks corresponding to both sets of the received pulses appear simultaneously on the screen, thus allowing the time delays to be measured. The acoustic speeds can then be calculated.

The attenuation coefficient is measured with the parallel surfaces of the sample directed at right angles to the beam. The amplitudes of the received pulses, either direct or reflected, firstly with the sample present and then removed are then compared with one another; the effect of the attenuation in the water is thus eliminated. Because the attenuation varies with frequency, the measurements are made with the aid of a spectrum analyser; with the use of heavily damped probes, attenuation coefficients can be determined for a wide range of frequencies without the need for probe changes (equation 2.10). Complications arising from oblique reflections and mode conversions within the block make it difficult for attenuation to be measured for finite angles of rotation. The method can be used with a range of temperatures by varying the temperature of the bath.

Art and Smith (1978) have described the on-line inspection of plastic pipes for wall-thinning when the pipes are rotated about their axes when immersed in water. The wall thicknesses are measured by measuring the time delay between the echoes from the inner and outer surfaces using the pulse method. The outer diameter is determined by rotating the pipe between two probes located one at each end of a line intersecting the axis. The probes both generate and receive pulses simultaneously and the delays t_1 and t_2 between the echoes received from the respective nearest outer walls are measured. If t is the transit time for direct propagation between the probes in the absence of a pipe and c is the speed of sound in water at the testing temperature, the outer diameter d of the pipe is given by:

$$d = c/[t - (t_1 + t_2)]. \tag{10.8}$$

Prassianakis and Theotokoglou (1992) report the use of attenuation measurements using the pulse-echo technique for assessing the damage to polymers under stress (section 5.3.3). They conducted the measurements on Perspex specimens, either with or without an edge crack or a V-shaped notch, as they were loaded progressively until fracture. Prassiankasis (1990) conducted both velocity and attenuation measurements on both undercured and properly cured samples of plasticized epoxy resins. It was found that changes in attenuation coefficient took place more rapidly than those in velocity with increase in percentage content of plasticizer.

10.4.3 The testing of rubber tyres

Although it may not always be the practice of tyre manufacturers to conduct 100% inspections of their products, ultrasonic tests are often made during the early stages of production runs to detect, and subsequently correct, any defects. Likely faults in tyres include the tendency for tread separation, non-uniformity of thickness of the casing and porosity in the actual rubber. The characteristic impedance of rubber is very close to that of water and, with the immersion technique, coupling losses are very small. On the other hand, the front- and back-wall echo amplitudes are very low. Hence, the through-transmission method is, in principle, more suitable for investigating defects but no real difficulties should arise with the use of the pulse-echo technique provided that the sensitivity of the flaw detector is high and focused probes are used.

The ultrasonic testing of tyres is well established and an early paper by Patton and Hatfield (1952) describes work making use of continuous waves at frequencies ranging from 50 to 250 kHz. Here, the tyre is rotated an a vertical plane about its axis in such a way that, at any time, a complete cross-section is completely immersed in water. The waves are radiated from a large transmitter with a divergent beam of 120° and six smaller receivers are located in the positions shown in Fig. 10.10. Reduction in amplitude, or disappearance, of the received signals is presumed to be indicative of the presence of a defect. More recently, a pulse-echo method (Halsey, 1968) using two 1 MHz focused PZT reversible probes has been developed. This again is a water-immersion method with the probes located externally to the carcass and mounted on a rigid support which could be displaced horizontally (Fig. 10.11). A complete scan can be made by suitably rotating the tyre which is inflated to working pressure. The sensitivity of the detector is sufficient to register echoes from the inner surface of the tyre. The

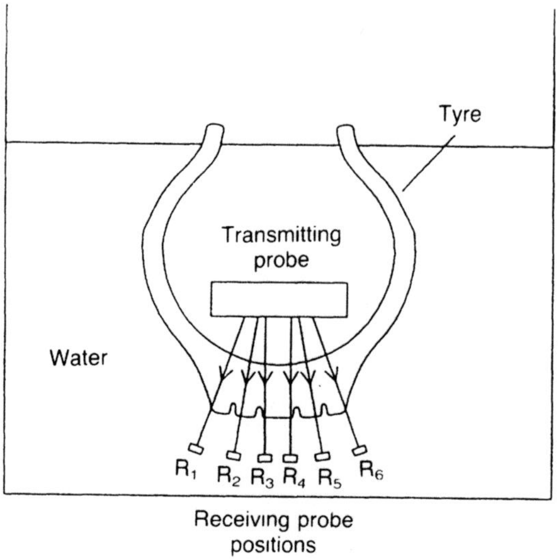

Fig. 10.10 Transmission technique for testing tyres (after Patton and Hatfield 1952)

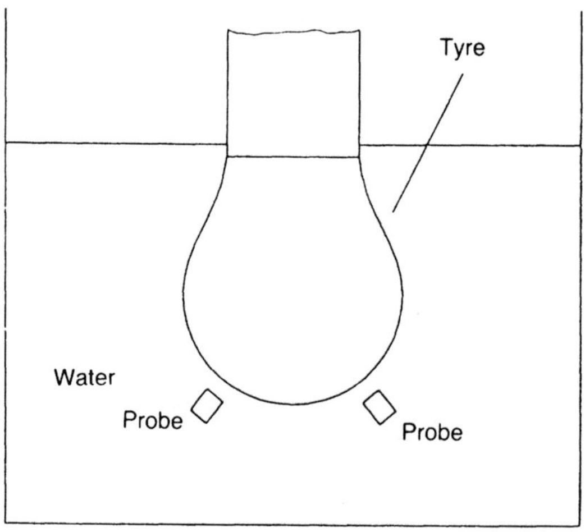

Fig. 10.11 Pulse-echo method for testing tyres (after Halsey 1968)

resolution is sufficiently high for detecting porosity and gross defects in the tread stock and at the interface between the tread and carcass.

10.5 THE TESTING OF FIBRE-REINFORCED PLASTICS

10.5.1 General Considerations

Fibre-reinforced plastics consist of strong uniaxial fibres, usually a few micrometres in diameter, having a high elastic modulus and embedded in an epoxy resin of low elastic modulus to form thin sheets of from 0.125 to 0.25 mm thick. They are bonded together to provide multilaminates, typically from 20 to 30 mm thick, similar to plywood and usually with the fibres in adjoining laminates orientated in different directions. Their densities are much lower than those of light alloys and they possess high tensile strengths which are comparable to high quality steel. For this reason, they are very suitable for the manufacture of structural components such as those used for aircraft. At present, the most important of these materials are carbon-fibre reinforced plastics (CFRP) and glass-fibre reinforced plastics (GRP), although boron-fibre reinforced plastics have been widely used in the past.

Because fibre-reinforced plastics are anisotropic materials, the speed of ultrasound in them depends on the direction of propagation. Reynolds and Wilkinson (1978) have shown that the speed of compression waves in single laminates of CFRP is about 12 000 m s^{-1} in the direction parallel to that of the fibres and decreases to about 2000 m s^{-1} as the angle between the directions of the ultrasound and the fibres increases to 90°. These figures are dependent on the respective speeds in the carbon and the resin in the same way as the those in steel and concrete for reinforced concrete (section 10.2.4). It would thus be expected that the speed of shear waves should decrease in the same way from 1750 m s^{-1} to 1250 m s^{-1} but this is not so. Reynolds and Wilkinson have also shown that the shear-wave speed increases with angle of orientation from 1750 m s^{-1} in the direction of the fibres to a maximum at an angle of 75° and then decreases to 1250 m s^{-1} when it is 90°. This apparent anomaly is attributed to mode conversion at the resin-fibre interfaces; the compression-wave speed in the fibres is nearly ten times greater than the shear-wave speed in the resin so that the critical angle (section 2.5.2) is about only 6°. Thus a great deal of the sound energy would be travelling though the fibres as compression waves before conversion back to shear waves for orientations greater than this angle.

Attenuation in fibre-reinforced plastics is caused mainly by inhomogeneity, e.g. porosity in the resin, and by deficiencies in the bonding of multilaminate samples. At high frequencies, attenuation may result from scattering by the fibres and by variations in the volume fraction, i.e. the ratio of the volume of the fibre to that of the resin, which is approximately 60% for optimum structural properties.

10.5.2 Defects in fibre-reinforced plastics

Any characteristic which may detract from the mechanical strength of a fibre-reinforced composite must be regarded as being a defect. An overall review of defect types, together with the appropriate non-destructive testing techniques, in fibre-reinforced composites (and adhesive joints) is given by Adams and Cawley (1988). Accounts are provided by Stone (1978) of defects in carbon-fibre reinforced plastics and by Golia *et al.* (1977) of those in glass-fibre reinforced plastics. Defects may arise from imperfections in the fabrication and in the bounding of the laminates and as a result of in-service and environmental damage. Defects in CFRP fabrications listed by Stone include void content and porosity, imperfect state of cure of resin, poor condition of the fibre-matrix interface, foreign inclusions and lack of uniformity of the fibre distribution. Additional defects can occur in CFRP samples made up from sheets of 'pre-preg': sheets of partially cured resin in which the fibres are uniformly distributed and orientated. They include translaminar cracks, delaminations, fibre misalignment within a sheet, non-uniformity of orientation and lay-up order of the plies, faults in end-butt joints of the sheets and overlap of and lack of side-butt within the plies. Similar defects may be found in other kinds of fibre-reinforced plastics.

Because composite materials are frequently joined with adhesive bonds, defects can occur in the bound lines. Causes of imperfect bonding include the incorrect shaping of the adherents (giving rise to variations in the gaps between them to be filled with adhesive), incorrect surface preparation of the plies, lack of homogeneity of the adhesives (resulting in the appearance of gaps), inclusions (or excess adhesive layers), faults in the conditions of the adhesives and interface conditions which may include the presence of unremoved release paper and moisture. In-service defects may arise from static overload and fatigue, degradation of the properties of the material and bonding, due to the presence of moisture, and damage.

10.5.3 Methods of testing

Testing for defects in fibre-reinforced plastics has been carried out with the ultrasonic pulse-echo and pulse-transmission methods and with the use of ultrasonic spectroscopy. Fibre-reinforced plastic materials generally exist in the form of large samples for which mechanical scanning is desirable. Ideally both sample and probe should be immersed in water but this may not always be feasible and, instead, coupling may be achieved by means of water jets.

For detecting easily resolvable defects such as delaminations, an A-scan (section 5.6) can be used. For very thin samples the use of short pulses is essential and frequencies of up to 25 MHz may be required. The degree of resolution in measuring the depths of delaminations is improved by means of weakly focused probes. However, A-scanning is unsuitable for detecting most defects which may occur in these materials and the use of a pulse through-transmission method with C-scanning is then necessary. The latter essentially measures the degree of attenuation which depends on the presence of delaminations, inclusions, internal damage occurring during service, void content and fibre density. Attenuation is also caused by reflection and transmission losses at the surfaces but it is shown by Jones and Stone (1976) that, provided that the surfaces are of even texture and not too greatly curved, the surface effects can be eliminated by measurements on two or more samples having identical surface characteristics but different thicknesses.

Difficulties may arise with the use of focused probes because of the existence of comparatively large variations of attenuation along the axis of the beam. Variations in the surface contours of the sample cause corresponding changes in distance from the probe. This affects the amount of energy transmitted and, hence, the measured attenuation in the sample. These difficulties may be overcome to a great extent by reducing the cross-section of the beam by placing a stop in front of the surface of the probe. Jones and Stone used laminated plastic stops to produce apertures having diameters of 2–7 mm.

The simplest arrangement for the through-transmission method is to locate separate transmitting and receiving probes on either side of the sample and maintained at a fixed distance apart with the aid of a calliper device. When access from only one side is possible, a single transmitter-receiver probe may be used and the amplitude of the first back-wall echo observed. An alternative method, described by Stone (1978) is to place a glass plate parallel with and below the lower surface of the sample and to observe the first echo from the surface of the plate. This avoids uncertainties produced by reflection from the surfaces of the sample. Whichever technique is employed, the attainment of a permanent record is simplified by using a gating device which allows only the wanted signal to be observed. Stone and Clarke (1975) made use of a quantized display on paper, using the C-scan, whereby three different attenuation levels (4 to 8, 8 to 12 and 12 to 16 dB) are indicated by different shades, black, grey and white, respectively. In this way, a scaled visual image of te sample in which defective areas are shown in grey and white shades is obtained.

Difficulties due to contamination by couplants can be overcome with the use of air coupling with the use of a frequency sufficiently low to enable detectable transmission to take place (equation 2.46) but at the cost of resolution. Golis *et al.* (1977) achieved this by using a through-transmission method with C-scanning to examine glass-fibre reinforced plastic pipes. 40 kHz bimorph transducers (section 4.3.3) are located by means of a calliper device with the transmitter outside the pipe wall and the receiver inside it. The receiver is embedded in a layer of sound absorbent material so that only directly transmitted pulses could be received. The method is found to be effective in detecting laminations, uncured and resin-starved regions and unbonded inclusions. To a lesser extent, it is possible to detect the presence of moisture. The degree of resolution is restricted to detecting defects of greater than 12 mm in extent.

Rose *et al.* (1973) report an ultrasonic spectroscopy method for inspecting composite tubes. A heavily-damped 10 MHz probe, having a bandwidth of 0–20 MHz, propagates pulses through the material. Echoes from the inner surface are gated and fed into a spectrum analyser, the output of which is displayed on an oscilloscope screen, with the time base converted using a Fourier transform (section 2.9) to provide indications of frequency. The display consists of a series of rounded peaks located at intervals of approximately 1 MHz, as determined by the reciprocal of the pulse width, approximately 1 μs, in the absence of defects. With the appearance of a flaw the number of detectable peaks is reduced and the nature of the observed spectrum can be related to the characteristics of the defect.

Markham (1976) found it possible to distinguish between delaminations and voids in carbon-fibre composites, also with the aid of a spectrum analyser. Standing waves are set up between a delamination and the front surface and the resulting resonances given rise to interference with the spectrum as indicated by sharp dips on the oscilloscope screen. The occurrence of voids gives rise to a rapid fall-off in the amplitude of the received signal with increase in frequency. The depth, d, of a delamination below the surface is related to the frequency, f, corresponding to each dip, by (section 2.6.5):

$$d = nc/2f. \tag{10.9}$$

where n is an integer denoting the order of resonance and c the ultrasonic speed. Using this method, it is possible to detect delaminations immediately below the surface whilst, with the conventional pulse-echo pulse technique, there may not be sufficient resolution to do so.

Adams *et al.* (1991) and Dewen and Cawley (1992) have used a similar method of detecting the presence of defects in glass-fibre blocks by shock excitation. The block is mounted on soft-foam blocks and is lightly tapped by a hammer to set up free vibrations. The detector, in the form of a heavily damped piezoelectric transducer (an accelerometer), is attached to the block by wax. The output, again, is fed to a spectrum analyser which indicates the fundamental and overtone frequency signals on an oscilloscope screen. Any change in the pattern of the corresponding peaks (of shape and frequency) from that observed for a defect-free sample is characteristic of any defects.

10.6 THE TESTING OF CERAMICS

Disregarding their uses in the manufacture of household and ornamental articles, ceramics are employed in three different areas: in the construction of porcelain electrical insulators, refractory materials and building commodities. Ultrasonic methods have been used successfully for testing all of these items for mechanical strength, porosity, quality of adhesion (for two or more components) and defects. The latter may comprise gas inclusions and laminations. Porcelain has a fine structure and thus has a comparatively low attenuation coefficient. Consequently, it should be possible to test them at megahertz frequencies with conventional ultrasonic flaw detectors. However, the other types of ceramics often have to be tested at lower frequencies with the type of equipment designed to test concrete (section 10.2). The speed of compression waves in the finished product often lies between 5000 and 6000 m s^{-1} although, as indicated below, speeds in some ceramics far exceed these values.

Porcelain electrical insulators used for high voltage applications require 100% testing because any failure may prove to be catastrophic. The basic requirements of a porcelain insulator are a high dielectric strength (the ability to withstand a high voltage without a failure of insulation), a high mechanical strength, a lack of porosity and an absence of structural defects.

Porosity arises as a result of the evaporation of the water content of the original clay during firing. In an investigation made in Poland (Filipczynski *et al.* 1966), it was found that the speed of ultrasound in unfired clay (kaolin) decreased from 1800 to less than 800 m s^{-1} when the water content is reduced from 16 to 9%. During the initial firing at a temperature of about 1000°C, the water evaporates completely, leaving porosity, and the ultrasonic speed is increased abruptly to about 2000 m s^{-1} because of a change in structure of the material to a crystalline form: felspar. On continued firing, the pores become partly filled with felspar by amounts depending on the firing temperature, the optimum value being about 1380°C corresponding to a maximum porosity of 4%. Figure 10.12 shows how the speed of ultrasound, porosity and mechanical strength vary with firing temperature with the speed reaching its maximum at this temperature. The figure also shows

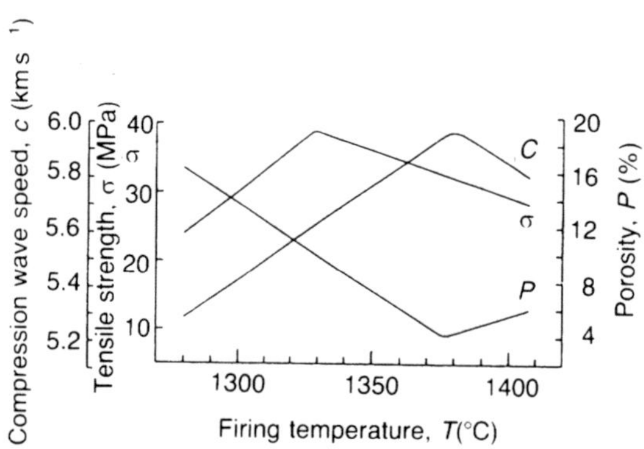

Fig. 10.12 Dependence on the speed, c, of compression waves, porosity P and tensile strength σ on the firing temperature, T, of electrically insulating porcelain (after Filipczynski, Pawlowski and Wehr 1966).

that the changes in speed lag behind those in mechanical strength. An increase in the strength of porcelain is also indicated by a decrease in the attenuation coefficient, which is to be expected in view of the increase in attenuation with porosity. More recent work at a frequency of 5 MHz has been reported by Panakkal (1992) who conducted measurements of both compression and shear waves to determine values with bulk and shear moduli and the Poisson ratio of clay ceramics, using equations 3.5 and 3.8. Values of each these quantities were shown to decrease with increase in pore volume fraction.

The ultrasonic testing of building and refractory ceramics has been discussed at some length by Davis and Brough (1972) who tested for mechanical strength, porosity and the presence of defects. Furthermore, they examined building materials, including tiles and bricks for moisture content, frost damage and quality of adhesion to backing surfaces. Because of the relatively short pulse lengths employed, they were able to test with a conventional ultrasonic flaw detector using 1 MHz probes using through transmission with both B- and C-scans.

Difficulties arise with coupling when conventional probes are used but Davis and Brough overcame them with the use of probes covered with either synthetic rubber or resin tips. These tips are sufficiently resilient for good contact to be made on rough surfaces by pressure alone so that it is unnecessary to use any couplant. However, for mechanical scanning, it is essential that this pressure remains constant at both opposite surfaces and the probes have to be subjected to constant spring-loading.

For ceramics having high compression-wave speeds, e.g. 12 000 m s^{-1} for which the frequency corresponding to a wavelength of 1 mm is 12 MHz, difficulties may arise when trying to locate small defects with the use of conventional piezoelectric transducers if attempts are made to cut them to resonance thicknesses (equation 4.6). This problem may be overcome with the use of piezoelectric plastic film transducers for which an upper frequency limit of 100 MHz is reported (section 4.3.4)

10.7 THE TESTING OF ADHESIVE BONDS

10.7.1 General considerations

Ultrasonic methods can be used for testing bonds between metal and metal, metal and non-metal, and non-metal and non-metal components. The types of bond considered here are those in laminates, overlap, honeycomb, sandwich, steplap and shrinkage joints (Fig. 10.13); the testing of laminates of fibre-reinforced plastics is discussed in section 10.5. Although shrinkage joints should not contain any adhesives, they are listed here because they are tested in the same way as other types of joints. The requirements for testing include the ability to detect the presence or absence of a bond and to assess the cohesive strength of an adhesive. Both pulse-echo and pulse through-transmission techniques, which are adaptable to A- and C-scanning, and also the Lamb-wave method (section 3.5) can be used. Two methods have been specially developed for this purpose: the Fokker bond tester and the ultrasonic impulse-induced resonance method.

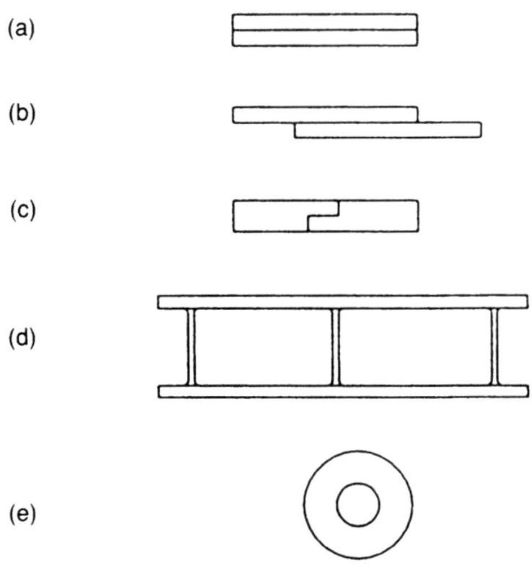

Fig. 10.13 Types of bonded joints: (a) laminate; (b) overlap; (c) step-lap; (d) honeycomb; (e) shrinkage.

When testing for lack of bonding, it is essential to consider the possibility of any liquid entering the space intended for the adhesive and thus giving rise to transmission, e.g. when oil seeps into a shrinkage joint in a bearing assembly.

A review of NDT techniques for testing bonded joints (and also composites) is given by Adams and Cawley (1988).

10.7.2 The pulse-echo technique

The simplest method of detecting for lack of bonding is the pulse-echo technique. This is effective when the thicknesses of the components to be bonded together are large compared with the wavelength, in order that the echoes from each interface can be clearly resolved from one another. The method has been used most successfully for the testing of metal-to-metal bonds but it is less satisfactory with non-metals because of their higher attenuation coefficients. However, Jones and Stone (section 10.5.3) found it possible to detect delamination in carbon-fibre reinforced plastics with this method by using focused high-frequency probes. Rose and Meyer (1973) used an immersion technique to test, at a frequency of about 10 MHz, the quality of an aluminium-to-aluminium bond by measuring the ratio of the amplitude of the front-wall echo to that of the bond echo. This ratio was found to increase with the bond strength as determined by the failure load provided by a mechanical testing machine which subjected the joint to shear stresses. Rose and Ditri (1992) have demonstrated the feasibility of the use of pulse-echo and through-transmission Lamb waves for inspecting bonds between aluminium layers.

10.7.3 The pulse through-transmission technique

The through-transmission technique can be used for testing bonded joints in the same way as it is used for detecting delamination in fibre-reinforced plastics (section 10.5.3). The amplitude of the received signal is compared with a reference level, such as the amplitude of the transmission signal. Lack of adhesion in honeycomb structures can be detected in this way at comparatively low frequencies like 0.5 MHz.

A scanning device in the form of a search-wheel probe was developed by Schliekelmann (1972) for the testing of metal-to-metal bonds (Fig. 10.14). Probes are mounted in the hubs of wheels, filled with liquid, which are fitted with rubber tyres in contact with each side of the sample. As the wheels rotate, the probes remain stationary and lined up with respect to one another for the duration of the scan. Coupling is maintained by directing water jets on to the areas of the required contact between the tyres and the sample.

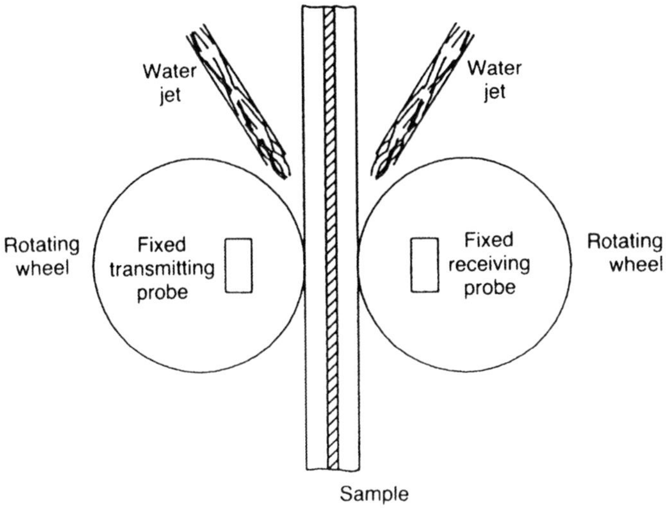

Fig. 10.14 Rotating search-wheel (or roller) probe (after Schliekelmann 1972).

10.7.4 Lamb wave measurements

The use of Lamb waves (sections 3.5 and 5.10) is especially suitable for testing a bond between a thin layer of a material and a thicker one, such as a 0.1 mm thick aluminium foil glued to a Perspex plate. Using a conventional flaw detector, the sample is immersed in a liquid together with a pair of probes directed to the surface at an angle (Fig. 10.15). A barrier, located as shown, prevents reflections at the surface from reaching the receiving probe. Scanning is achieved by moving the sample horizontally whilst the probes remain stationary. For good adhesion, no Lamb waves are observed but with lack of adhesion, a received signal appears on the oscilloscope screen. The required values of probe angle and frequency are governed by the speed of sound in the water and the thickness of the foil (section 3.5). In practice, these quantities can be determined empirically by deliberately introducing a disbond in a model sample. As an alternative to using immersion, a pair of identical shear-wave probes, of appropriate angle and located at a fixed distance from each other, are scanned in tandem across the surface of the thin layer, using a conventional couplant.

10.7.5 The Fokker bond tester

The Fokker bond tester has been shown to be highly successful in testing epoxy resin metal-to-metal bonds and also metal-to-honeycomb bonds (Smith and Cagle, 1966). It measures changes in impedance and

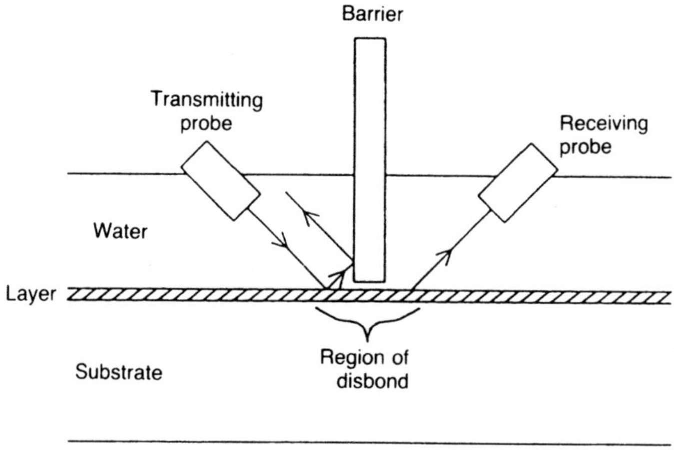

Fig. 10.15 Testing for bonding of thin layers on thicker substrates with Lamb waves (after Worlton 1957)

resonance frequency to determine the quality of a bond. The probe consists of a disc-shaped transducer which is spring-loaded on to the top of the sample and coupled to it by means of a viscous substance, e.g. grease or thick detergent. The mounting of the transducer is such that it can resonate in both thickness (or axial) and radial modes and thus propagate both compression and shear waves at the respective resonance frequencies. The choice of the transducer is governed by the thickness of the sample to be tested, but the fundamental resonance frequencies are always less than 1 MHz for thickness modes and 0.5 MHz for radial modes.

The transducer is driven at a constant voltage over a wide frequency band and the resonances, both thickness and radial, are each indicated by sharp dips on the cathode-ray screen connected to a spectrum analyser. The amplitude of resonance is indicated by a microammeter. Two resonances are generally observed, i.e. the thickness and radial for a given frequency mode, the latter usually being at approximately half the frequency of that of the former. The probe, when placed in contact with a sheet of metal, suffers a change in impedance and, consequently there is a change in both the resonance frequency and amplitude. The corresponding shifts are indicated by the oscilloscope and the microammeter and can be related to the thickness and nature of the matal.

The choice of mode of vibration depends on the kind of joint to be examined. Bonds such as the overlap type, which are liable to suffer shear stresses when in service, are examined in the radial mode so that shear stresses are introduced to the adhesive layer. On the other hand, honeycomb joints, which are normally subjected to tensile stresses, are tested in the thickness mode. In each case, the probe is initially coupled to a metal test piece made of the material forming the upper member of the bond and having the same thickness. A shift in both amplitude and frequency for the selected resonance mode is observed. The dip indicating the resonance frequency is adjusted horizontally to appear at the centre of the oscilloscope screen and the amplitude level is adjusted to give a convenient reading of the microammeter, e.g. 100 scale divisions.

When testing a metal-to-metal joint for bond strength, the probe is coupled to the surface of the sample to be tested and, provided that there is no disbond, shifts of both frequency and amplitude for radial resonance are observed. These changes are caused by the additional loading provided by the lower member of the bond and by the attenuation in the adherent. The latter varies with bond strength and, in practice, the instrument is initially calibrated by measuring bonds having different strengths which are later measured by testing to destruction.

When testing a honeycomb joint, with the thickness (or axial) mode, there is no shift in frequency because of the very low loading of the honeycomb compared with that of the metal sheet, but the effect of the bond is always to decrease the amplitude by an amount depending on its quality due to the damping effect of the honeycomb on the metal sheet.

10.7.6 The ultrasonic impulse-induced resonance method

The ultrasonic impulse-induced resonance method (Papadakis, 1978) is a form of pulse-resonance technique (section 5.5) designed to test bonds between metal sheets and non-metals with a conventional flaw detector. Wide-band pulses, long enough to induce stationary waves, are propagated through a Perspex delay medium into the metal sheet, which has thickness resonance frequencies (section 2.6.5) within the bandwidth of the pulses. The received signals are left unrectified and the first observed echo consists of the pulses reflected at the top surface of the sheet, followed by a 'ringing tail' caused by the stationary waves in the metal. A portion of this ringing tail is gated and fed into a spectrum analyser which indicates the frequencies of the fundamental and harmonics of the half-wavelength thickness resonances.

When the non-metallic sheet is bonded to the underside of the metal sheet, the resonances are damped by amounts depending on the degree of bonding. Thus a disbond gives high amplitudes at resonance, a partial bond provides moderate amplitude and a perfect bond provides a low amplitude. Papadakis experimented successfully with metal sheets ranging in thickness from 0.2 to 4 mm and bonded to highly attenuating non-metals, including cardboard, rubber and an asbestos-filled paper. With wide-band 2.5 and 5 MHz probes, resonances of up to the 5th harmonic were obtained. With a thermosetting adhesive, an amplitude difference of 12 dB could be observed between a perfect bond and a perfect disbond.

REFERENCES

Adams, R.D., Brownjohn, J.M.W. and Cawley, P. (1991) *NDT Internat*, **24**, 123–34.

Adams, R.D. and Cawley, P. (1988) *NDT Internat*, **21**, 208–22.

Ast, J.D. and Smith, J.G. (1978) *Plast Eng*, **34**, 41–4.

BS 1881 (1986) *Recommendations for Nondestructive Methods of Test for Concrete*, Part 203: *Measurements of the Velocity of Ultrasonic Pulses in Concrete*; Part 201: *Guide to the use of Nondestructive Methods of Tests for Hardened Concrete*. British Standards Institution, London.

Bucur, V. (1983) *NDT Internat*, **21**, 116–26.

Davis, W.R. and Brough, R. (1972) *Ultrasonics*, **10**, 118–26.

Dewen, P.N. and Cawley, P. (1992) *NDT Internat*, **25**, 66–75.

Elvery, R.H. and Nwokoye, D.N. (1970) *Proc Symp Non-Destr Test Concrete and Timber 1969* Institute of Civil Engineers, London.

Elvery, R.H. (1976) *Mag Concrete Res*, **28**, 181–90.

Filipczynski, L., Pawlowski, Z. and Wehr, J. (1966) *Ultrasonic Methods of Testing Materials*, Butterworths, London.

Golis, R.J., Grove, J.C., Gaspero, N.A. and Medrano, J. (1977) *ASTM Spec. Tech. Pub. 521*. American Society for Testing and Materials, 311–25.

Halsey, G.H. (1968) *Mater Eval*, **26**, 137–42.

Jones, B.R. and Stone, D.E.W. (1976) *NonDestr Test*, **9**, 71–4.

Jones, R. (1962) *Non-Destructive Testing of Concrete*. Cambridge University Press, Cambridge.

Makow, D.M. (1969) *NonDestr Test*, **2**, 197–9.

Malhorra, V.M. and Carino, N.J. (1991) *Handbook on Nondestructive Testing of Concrete*. CRC Press Inc., Boston.

Markham, M.F. (1976) *Plastics and Rubber: Materials and Applications*, **1**, 187–92.

Nwokoye, D.N. (1971) *Nondestructive Assessment of Shear Wave Strength of Tubes* Tech. Note 27. Construction Industry Research and Information Association (CIRIA), London.

Panakkal, J.P. (1992) *Br J NDT*, **34**, 529–32.

Papadakis, E.P. (1978) *Mater Eval*, **36**, 37–40.

Patton, R.G. and Hatfield, P. (1952) *Electron Ind*, **24**, 522–5.

Prassianakis, J.N. (1990) *J Appl Polym Sci*, **39**, 2031–41.

Prassianakis, J.N. and Theotokoglou, E.N. (1992) *Eng Fract Mechanics*, **43**, 117–21.

Read, B.E. and Dean, G.D. (1978) *The Determination of Dynamic Properties of Polymers and Composites*. Adam Hilger, Bristol.

Reynolds, W.N. and Wilkinson, S.J. (1978) *Ultrasonics*, **16**, 159–63.

Rose, J.L., Carson, J.M. and Leidel, D.J. (1973) *Spec. Tech. Pub. 521*. American Society for Testing and Materials, 311–25.

Rose, J.L. and Ditri, J.J. (1992) *Br J NDT*, **34**, 591–4.

Rose, J.L. and Meyer, P.A. (1973) *Mater Eval*, **31**, 109–14.
Schliekelmann, R.J. (1972) *NonDestr Test*, **5**, 144–53.
Smith, D.F. and Cagle, C.V. (1966) *Mater Eval*, **24**, 362–70.
Stone, D.E.W. (1978) *Br J NDT*, **20**, 63–75.
Stone, D.E.W. and Clarke, B. (1975) *NonDestr Test*, **9**, 71–4.
Vold, R.C. and Hope, B.R. (1978) *Br J NDT*, **20**, 232–41.
Whitehurst, E.A. (1951) *J Am Concrete Inst*, **22** (Proceedings 47), 433–44.

CHAPTER 11

Training, certification and standards

11.1 TRAINING AND CERTIFICATION

In the early days of ultrasonic testing, any examinations were carried out in an *ad hoc* basis and the operator would have to rely mainly on the printed instructions issued by the manufacturer of the equipment concerned. However, with advancements in sophistication in the design of ultrasonic equipment and stricter specifications for testing, the instructions became more complex and it would then have been necessary for any personnel concerned with the use of this equipment to attend a course run by either the manufacturer or his agent.

An important consideration in connection with ultrasonic testing is that, for manual operations, there are no permanent records of any tests, in contrast with radiography for which a record in the form of a radiograph is obtained. For this reason, it is vital that the operator has received sufficient training and experience and has been subjected to a thorough examination of their ability. However, it is not possible to test their integrity in any examination and it is thus essential that the employer keeps a watchful eye on the quality of an operator in this respect.

Training in the use of ultrasonic instruments and in the methods of examination of products such as forgings, castings, welds and aerospace components are essential requirements for an operator and expertise in this field is essential if certification is required. All certification schemes include the following of relevant courses for the ultrasonic testing of components at various levels for which detailed syllabuses together with the appropriate number of hours of study are laid down.

In the UK, approval of the establishments providing training in ultrasonic and other methods of testing is given by the British Institute of Non-destructive Testing, which publishes a list of those which have received their approval. The listed training establishments are required to employ teaching staff at the correct level of competence and to issue suitable instruction notes to be followed by the trainees. They should possess adequate supplies of ultrasonic equipment, probes, calibration blocks and test specimens. An annual audit is carried out by the British Institute of Non-destructive Testing to ensure that standards are properly maintained. These approved establishments may act as centres for certification examinations set for Personnel Certification in Non-destructive testing (PCN) qualifications, as discussed later. The training and examination facilities must be separated to comply with PCN regulations. This 'central certification' has the advantage that the examination is independent of outside influences such as the trainees' employers. The attendance at one of these establishments is thus essential if a PCN qualification is required.

In the USA, the qualifying body is the American Society of Nondestructive Testing (ASNT) which has produced a document SNT-TC-1A which has been used since 1968 as a guideline for employers in training and examining their personnel. In 1991, an new standard ASNT-CT-189 was introduced. This differs from the earlier document in that, where appropriate, the word 'shall' replaces the word 'should' so as to remove the employers' option not to comply with the Society's regulations. In addition, aerospace workers are now obliged to work to another variation of the original document: MIL-STD-410E.

For the purpose of establishment of a training programme, the employer may choose from this document suitable items from the printed syllabus and he should comply with the procedures regarding the minimum periods of training and experience. Here the training is supervised and examined by a suitable qualified (Level III, as explained below) supervisor on behalf of the employer.

Certificated non-destructive testing personnel are classified into three levels as results of the examinations mentioned above: *Level I* operator, *Level II* technician, *Level III* supervisor. By definition, a Level I operator must be able to prove his ability to conduct basic tasks such as calibration and working to a set procedure in a satisfactory manner under the supervision of a Level II NDT technician. The latter should be able to conduct any ultrasonic procedure and testing for which he is qualified, e.g. to set standards of sensitivity and to size flaws. He must also be responsible for the work carried out by the Level I operator. When in any serious doubt, he should seek advice from a Level III supervisor who takes full responsibility for the work of his subordinates. Before taking the examination, the candidate must possess a valid eyesight test certificate.

The periods of validity of the certificates are 5 years if awarded by PCN but only 3 years for those awarded by ASNT.

There follows a selection of standards applicable to education and training in ultrasonic testing.

European

EN 45013 1989 General Criteria for Certification Bodies operating Certification of Personnel.

EN 473 1993 Qualification and Certification of NDT Personnel.

EN 4179 1995 Personnel Qualification and Certification, NDT Aerospace Series.

USA

SNT-TC-1A 1992 Personnel Qualifications and Certification of NDT Personnel.

ASNT-CP-189 1991 Standard for Qualification and Certification of NDT Personnel.

MIL-STD-410E 1991 NDT Personnel Qualification and Certification.

API RP 2X 1998 Guidelines for Qualification of Ultrasonic Technicians.

International

ISO 9712 1992 Nondestructive Testing – Qualifications and Certification of Personnel.

11.2 STANDARDS

Standards for ultrasonic testing have been prepared in all of the major developed countries and, although they are published in the language of the country concerned, English translations are sometimes available. A selection of titles of these standards, which are restricted to those either written in or translated into English, is listed below and dates, where shown, are those known by the authors at the time of writing. Addresses of the standards organizations quoted below are given in Appendix G.

In Europe, there is a major reorganization is standards to achieve uniformity. This is being carried out by the Comité Européen de Normalisation (CEN) of which the major European countries, including the UK which is represented by the British Standards Institution (BSI), are members. This reorganization will extend over a long period and completion may not be achieved before the end of the twentieth century.

Another organization responsible for setting standards and codes of good practice in ultrasonic testing is the International Standards Organisation (ISO), which extends worldwide and very good relationships exist between the BSI, CEN and ISO. The following standards are in common use for ultrasonic testing.

British

BS 1134 1988 Assessment of Surface Texture.

BS 1881 1986 Recommendations for Non-Destructive Methods of Tests for Concrete.

Part 201: Guide to the use of Nondestructive Methods of Tests for Hardened Concrete.

Part 203: Measurements of the Velocity of Ultrasonic Pulses in Concrete.

BS 2704 1978, reconfirmed 1983 Specifications for Calibration Blocks for use in Ultrasonic Flaw Detection.

BS 3683 1985 Glossary of Terms used in Non-destructive Testing. Part 4: Ultrasonic Testing.

BS 3889 Non-Destructive Testing of Pipes and Tubes.

Part 1: 1991 Methods of Automatic Ultrasonic Testing for the Detection of Imperfections in Wrought Steel Tubes.

BS 3923 Methods of Ultrasonic Examination of Welds.

Part 1: 1986 Manual Examination of Fusion Welds in Ferritic Steels.

Part 2: 1972 Automatic Examination of Fusion Welded Butt Joints in Ferritic Steels.

BS 4124 1991 Methods for Ultrasonic Detection of Imperfections in Steel Forgings.

BS 4331 Methods for Assessing the Performance Characteristics of Ultrasonic Flaw Detection Equipment.

Part 1: 1978 reconfirmed 1983 Overall Performance: On-Site Methods.

Part 2: 1972 reconfirmed 1982 Electrical Performance.

Part 3: 1974 reconfirmed 1982 Guidance of the In-service Monitoring of Probes (Excluding Immersion Probes).

BS 5996 1980 Methods for Ultrasonic Testing and Specifying Quality Grades of Ferritic Steel Plates.

BS 6208 1990 Methods for Ultrasonic Testing and Specifying Quality Levels of Ferritic Steel Castings.

BS 7706 1993 Methods for Ultrasonic Time of Flight Diffraction.

German

DIN 54120 1966, 1973 Zerstorungsfreie Prüfung: Kontrollkorper 1. (English translation available from the British Standards Institution).

USA

In the USA, the standards authority is the American Society for Testing and Materials (ASTM). The standards cited below are published yearly, updated where necessary. For this reason, dates are omitted.

ASTM A 21 Standard Specification for Carbon Steel Axles, Non Heat-Treated and Heat-Treated, for Railway Use.

ASTM A 388 Standard Practice for Ultrasonic Examination of Heavy Steel Forgings.

ASTM A 418 Standard Method of Ultrasonic Inspection of Turbine and Generator Steel Rotor Forgings.

ASTM A 435 Standard Specification for Straight-Beam Ultrasonic Examination of Steel Plates for Pressure Vessels.

ASTM A 503 Standard Specification for Ultrasonic Examination of Large Forged Crankshafts.

ASTM A 531 Standard Recommended Practice for Ultrasonic Examination of Turbine-Generator Steel Retaining Rings.

ASTM A 577 Standard Specification for Ultrasonic Angle-Beam Examination of Steel Plates.

ASTM A 578 Standard Specification for Straight-Beam Ultrasonic Examination of Plain and Clad Steel Plates for Special Applications.

ASTM A 609 Standard Specification for Ultrasonic Examination of Carbon and Low-Alloy Steel Castings.

ASTM A 745 Standard Recommended Practice for Ultrasonic Examination of Austenitic Steel Forgings.

ASTM B 548 Standard Method and Specification for Ultrasonic Inspection of Aluminium Alloy Plate for Pressure Vessels.

ASTM B 594 Standard Method for Ultrasonic Inspection of Aluminium Alloy Wrought Products for Aerospace Application.

ASTM D 2845 Standard Method for Laboratory Determinations of Ultrasonic Pulse Velocities and Elastic Constants of Rock.

ASTM E 113 Ultrasonic Testing by the Resonance Method.

ASTM E 114 Ultrasonic Pulse-Echo Straight-Beam Testing by the Contact Method.

ASTM E 127 Fabricating and Checking Aluminium Alloy Ultrasonic Standard Reference Blocks.

ASTM E 164 Ultrasonic Contact Examination of Weldments.

ASTM E 213 Ultrasonic Inspection of Metal Pipe and Tubing.

ASTM E 214 Immersed Ultrasonic Testing by the Reflection Method using Pulsed Longitudinal Waves.

ASTM E 273 Procedures for Ultrasonic Inspection of Longitudinal and Spiral Welds of Welded Pipe and Tubing.

ASTM E 317 Standard Practice for Evaluating Performance Characteristics of Ultrasonic Pulse-Echo Testing Equipment without the Use of Electronic Measurement Instruments.

ASTM E 428 Fabrication and Control of Steel Reference Blocks.

ASTM E 494 Measuring Ultrasonic Velocity in Materials.

ASTM E 500 Definitions of Terms Relating to Ultrasonic Testing.

ASTM E 587 Ultrasonic Angle Beam Examination by the Contact Method.

ASTM E 588 Detection of Large Inclusions of Bearing Quality Steel by the Ultrasonic Method.

ASTM E 664 Measurement of the Apparent Attenuation of Longitudinal Ultrasonic Waves by the Immersion Method.

ASTM E 797 Measuring Thickness by Manual Pulse-Echo Contact Method.

ASTM E 804 Ultrasonic Test System by Extrapolation between Flat Bottom Hole Sizes.

International (ISO) and European (CEN) standards

In many instances, ISO and CEN standards are identical. Here, the ISO number is given first and, where appropriate, the CEN number is bracketed.

ISO 2400 1972 Welds in Steel: Reference Block for the Calibration for Ultrasonic Examination.

ISO 5817 (CEN 25817 1992) Arc-Welded Joints in Steel – Guidance on Quality Levels for Imperfections.

ISO 6250 (CEN 26250 1992) Classification of Imperfections in Metallic Fusion Welds, with Explanations.

ISO 7963 (CEN 27963 1992) Specification for Calibration Block No. 2 for Ultrasonic Examination.

ISO 9712 (CEN 473 1992) Qualification and Certification of NDT Personnel.

The following standards are in the process of preparation at the time of writing and completion is expected during 1995:

EN 583 Ultrasonic Examination.
Part 1 General Principles for Ultrasonic Testing.
Part 2 Sensitivity and Range Setting.
Part 3 Transmission Technique.
Part 4 Examination for Defects Perpendicular to the Surface.
Part 5 Characterisation and Sizing of Discontinuities.

EN 27963 Welds in Steel: Specification for Calibration Block No. 2 for Ultrasonic Calibration.

CEN/TC/138–10 Calibration Block No. 1 for Ultrasonic Calibration.

Work scheduled for completion beyond 1995 includes the following:

Ultrasonic Examination – Characterisation and Verification of Equipment for Ultrasonic Examination, Parts 1, 2 and 3.
Ultrasonic Thickness Measurement.
Automated Ultrasonic Examination.

Further reading

It is unfortunate that there is a dearth of textbooks on ultrasonic testing at the level of this book. Many works on this subject are directed either to the operator with minimal educational background or to experienced researchers. Others, however, contain more detailed information than can be accommodated in a book of this size. The following are highly recommended:

Filipzcynski, L., Pawlowski, Z. and Wehr, J. (1966) *Ultrasonic Methods of Testing Materials*. Butterworths, London. Out of print but copies available for reference in some libraries.

Krautkrämer, J. and Krauträmer, H. (1990) *Ultrasonic Testing*, 4th edn, Springer Verlag, Berlin. This book covers the subject in great detail.

McIntire, P. (ed) (1991) *Nondestructive Testing Handbook*, 2nd edn, Vol 7, *Ultrasonics*. American Society for Nondestructive Testing, Columbus, OH.

McMaster, R.C. (ed) (1963) *Nondestructive Testing Handbook* 1st edn, Vol 2, Ronald, New York. Now out of print but still available for reference in many appropriate libraries.

Szilard, J. (ed) (1982) *Ultrasonic Testing*. Wiley, London. This book contains contributions from a number of authors on important developments in ultrasonic testing at the time of its publication.

Current developments in ultrasonic testing can be followed by referring to the following journals:

Insight (formerly *The British Journal of Non-destructive Testing*), monthly, the British Institute of Non-destructive Testing, Northampton.

Materials Evaluation, monthly, American Society for Nondestructive Testing, Columbus, OH.

NDT & E International (formerly *NDT International*), bimonthly, Butterworth-Heinemann, Elsevier Science, Oxford.

Nondestructive Testing and Evaluation (formerly *Nondestructive Testing Communications*), 6 issues per volume, Gordon and Breach, London.

Ultrasonics, bimonthly, Butterworth-Heinemann, Elsevier Science, Oxford.

Detailed review articles covering the whole spectrum of non-destructive testing have appeared from time to time during the period 1970 to 1985 in the undermentioned publications for which contributions related to ultrasonic testing are listed.

Research Techniques in Nondestructive Testing, R.S, Sharpe ed., Academic Press, London.

Volume 1 (1970)

Acoustic Emission, P.H. Hutton and R.N. Ord.

Ultrasonic Spectroscopy, O.R. Gericke.

Ultrasonic Imaging Systems, J.E. Jacobs.

Ultrasonic Critical Angle Reflectivity, F.L. Becker and R.L. Richardson.

Ultrasonic Holography, E.E. Aldridge.

Volume 2 (1974)

Means of Visualising Ultrasound, D.M. Marsh.

The Reduction and Display of Ultrasonic Data, D.S. Dean and D. Young.

Ultrasonic Image Synthesis, L. Kay.

Electromagnetic Generation of Ultrasound, E.R. Dobbs.

Volume 3 (1977)

Ultrasonic Frequency Analysis, L. Adler, K.V. Cook and W.A. Simpson.

Sizing Crack-Like Defects by Ultrasonic Means, M.G. Silk.

Ultrasonic Correlation Techniques, V.L. Newhouse and E.S. Furgason.

Electronic Steering of Ultrasound for Tube Testing, K.R. Whittington.

Characterisation of Ultrasonic Transducers, I. Bredael.

Volume 4 (1980)

Possible Application of Inverse Scattering Concepts to Ultrasonic Imaging, R.H.T. Bates and G.C. McKinnon.

Structure Analysis by Scattered Ultrasonic Radiation, K. Goebbels.

Ultrasonic Measurements of Material Properties, A. Vary.

Applications of Surface Acoustic Wave (SAW) Devices in NDT, P. Das and D.R. Arsenault.

Ultrasonic Techniques for Inspecting Austenitic Welds, M.G. Silk.

Defect Characterisation using Ultrasonic Techniques, A. Rogerson and R.A. Murgatroyd.

Volume 5 (1982)

A Review of Large-angle Low-frequency Ultrasonic Scatter Analysis, D.S. Dean.

Caustics and the Inversion of Ultrasonic Scattering Data, P.A. Doyle, J. Latimer and L. Adler.

Laser Generation of Ultrasound in Metals, C.B. Scruby, R.L. Dewhurst, D.A. Hutchins and S.B. Palmer.

Volume 6 (1983)

Ultrasonic Synthetic-aperture Focusing Techniques in NDT, J. Seydel.

Ultrasonic Focusing Techniques, S.O. Harrold.
Methods of Computer Modelling of Ultrasonic Waves In Solids, L.J. Bond.
The Use of Ultrasonics to Measure Residual Stresses, D.R. Allen, W.H.B. Cooper, C.M. Sayers and M.G. Silk.

Volume 7 (1984)
Electron Beam Forming, Defect Reconstruction and Classification by Ultrasonic Phased Arrays, W.G. Gebhardt.
Mode-conversion Ultrasonic Testing, L.J. Bond and N. Saffari.

Volume 8 (1985)
Computer-controlled Systems for Ultrasonic Testing, P.J. Mudge.
The Application of Microprocessors in Ultrasonic Nondestructive Testing, R.A. Bacon and R.C. Chivers.
Ultransonic Focusing Techniques, R. Saglio and A.C. Prot.
Quantitative Acoustic Emission Techniques, C.B. Scruby.
Ultrasonic Phase Information, R.C. Chivers.

Free damped vibrations (section 2.2.2)

When the mass M is displaced in the axial direction of the spring by some amount y_0, so as to apply a tensile stress, and then released, the spring loses elastic energy to the mass which consequently gains kinetic energy: the energy of motion. At a given time, t, when the **displacement** of the mass from its rest position is equal to y, the total mechanical energy, E_m, in the system is equal to the sum of elastic energy $\frac{1}{2}y^2/C_m$ stored in the spring and the kinetic energy $\frac{1}{2}M(dy/dt)^2$ of the mass:

$$E_m = \tfrac{1}{2}M(dy/dt)^2 + \tfrac{1}{2}y^2/C_m. \qquad \text{(A.1)}$$

To simplify the analysis, it is assumed that the value of the mass of the spring is negligible compared with M and thus can be ignored. The energy, E_m, is dissipated in overcoming the frictional force, $R_m dy/dt$, between the mass and the surface and the resultant rate of change in energy is given by:

$$-dE_m/dt = R_m(dy/dt)^2 \qquad \text{(A.2)}$$

For a moving system where the solution $dy = 0$ for all values of y is invalid, equations A.1 and A.2 yield:

$$Md^2y/dt^2 + R_m dy/dt + y/C_m = 0. \qquad \text{(A.3)}$$

This is a second order differential equation and is similar in form to that representing the discharge of an electrical capacitor through an inductor in series with a resistor. Thus M, R_m, C_m and y are respectively analogous to the inductance, L, capacitance, C, resistance, R, and charge, Q. Using a method of induction, the solution to equation A.3 can be shown to be:

$$y = \exp(-\alpha' t)\{A \exp[\alpha'^2 - \omega_0{}^2]^{1/2}t + B \exp[-(\alpha' t^2 - \omega_0{}^2)^{1/2}t]\}, \qquad \text{(A.4)}$$

where $\alpha' = R_m/2M$ and $\omega_0 = (1/C_m M)^{1/2}$.
Here A and B are constants having values which can be determined by applying appropriate boundary conditions, as shown below. With

ultrasonic testing, we are interested in oscillations of the source which occur when α' is less than ω_0; equation A.4 is then modified to become:

$$y = \exp(-\alpha' t)[A\exp(j\omega t) + B\exp(-j\omega t)], \qquad (A.5)$$

where $j = (-1)^{1/2}$ and $\omega^2 = \omega_0{}^2 - \alpha'^2$.

The boundary conditions defined in section 2.2.2, when $t = 0$, are:

(a) $y = y_0$, so that $y_0 = A + B$, and
(b) $dy/dt = 0$, so that:

$$dy/dt = \exp(-\alpha' t)\{j\omega[A\exp(j\omega t) - B\exp(-j\omega t)] \\ -\alpha'[A\exp(j\omega t) + B\exp(-j\omega t)]\} = 0$$

i.e.:

$$A(j\omega - \alpha') = B\,(j\omega + \alpha')$$

from which:

$$A = (y_0/2)(1 - j\alpha'/\omega) \quad \text{and} \quad B = (y_0/2)(1 + j\alpha'/\omega)$$

Now:

$$\cos(\omega t) = [\exp(j\omega t) + \exp(-j\omega t)]/2$$

and

$$\sin(\omega t) = [\exp(j\omega t) - \exp(-j\omega t)]/2j.$$

Substituting the above expressions into equation A.5 gives:

$$y = y_0\exp(-\alpha' t)[\cos(\omega t) + (\alpha'/\omega)\sin(\omega t)]. \qquad (A.6)$$

Consider a right-angle triangle of vertical height α' and base ω. Let ω_0 be the hypotenuse and ϕ the angle between α' and ω. We then have:

$$\omega_0{}^2 = \alpha'^2 + \omega^2 \quad \text{and} \quad \phi = \tan^{-1}(\alpha'/\omega) \qquad (A.7)$$

and equation A.5 can thus be written as:

$$y = (y_0/\cos\phi)\exp(-\alpha' t)[\cos(\omega t)\cos\phi + \sin(\omega t)\sin\phi] \\ = (y_0(\omega_0/\omega))\exp(-\alpha' t)[\cos(\omega t) - \phi],$$

which is equation 2.1.

Time delay produced with mode conversion on internal reflection at the lateral surface of a long cylinder (section 5.3.2)

Consider (Fig. B.1) a ray of axially-directed diverging comperession waves in a solid cylinder, surrounded by air, and incident at a glancing angle at a point, O, on the surface. Mode conversion takes place on internal reflection at O; let OM and ON be the respective compression and shear waves with the velocities c_L and c_T (section 2.5.2). Let ON strike the opposite surface at the point N and OP and MN represent the normals at O and M. The glancing angle is assumed to be sufficiently close to 90° for point, M, to be regarded as being on the surface. At point N some of the shear-wave energy is converted back to compression wave energy and there will thus be a time delay δt at N between longitudinal waves arriving there either direct from the source or after reflection at P, without mode conversion, and those which have undergone reflection and mode conversion at O. Putting the

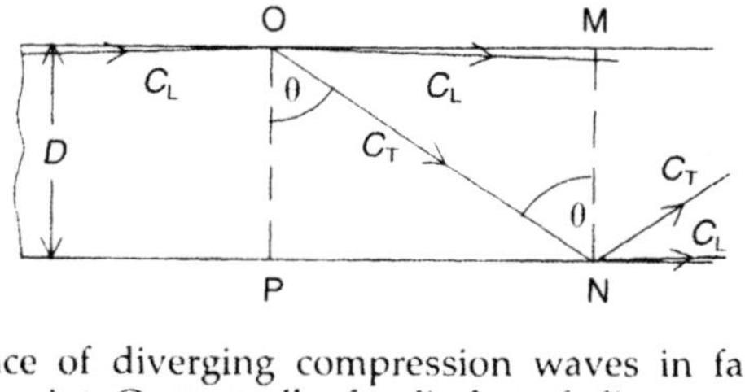

Fig. B.1 Incidence of diverging compression waves in far field at grazing angle (≈ 90°) at point O on wall of cylinder of diameter D with reflected compression waves, OM, at the same angle and speed c_L and reflected mode-converted shear waves ON at an angle θ and speed c_T.

diameter of the cylinder equal to D and the angles PON and MNO equal to θ we have:

$$\delta t = D/(c_T \cos\theta) - (D\tan\theta)/c_L.$$

Putting $\cos^2\theta = 1 - \sin^2\theta$ and $c_T/c_L = \sin\theta$ (equation 2.49(b)), it follows that:

$$\delta t = D(c_L{}^2 - c_L{}^2)^{1/2}/c_T c_L,$$

from which:

$$c_T = c_L/(1 + a^2)^{1/2},$$

where $a = c_L \delta t/D$.

Specific acoustic impedance for attenuated plane progressive waves (section 5.9)

In section 2.3.4 it is shown that the specific acoustic impedance, Z_a, of a material as defined by the ratio of the acoustic pressure, p, to the particle velocity, u, is entirely real for plane progressive non-attenuating plane waves and equal to the product ρc, where ρ is the density and c the speed of sound for the material. Here, by modifying the analysis given in that section, it is shown that the effect of attenuation is to add an imaginary component X_a to Z_a.

The particle displacement, y, for propagation in the x direction in an attenuating medium for which the attenuation coefficient is equal to α can be expressed as:

$$y = y_0 \exp(-\alpha x) \exp[j(\omega t - kx)],$$

where the meanings of the symbols are given in section 5.9. The particle velocity, u, is then given by:

$$u = \partial y/\partial t = j\omega y,$$

and, from equation 2.15, the acoustic pressure p by:

$$p = -\rho c^2(\partial y/\partial x) = \rho c^2(\alpha + jk)y.$$

Hence:

$$Z_a = p/u = \rho c(1 - j\alpha/k),$$

$$R_a = \rho c,\ X_a = -\rho c\alpha/k \text{ and } \tan\phi = X_a/R_a = -\alpha/k$$

where ϕ is the phase angle, remembering that $k = \omega c$.

Evaluations of the ratios of skip distance to sound path for pipes and plates having equal thicknesses for given ratios of diameter to wall thickness and probe angles (equations 7.4, 7.5 and 7.6)

A shear wave probe of angle α, located at point P on the surface of a pipe of diameter D and thickness t, propagates shear waves in the direction PR (Fig. D.1(a)), suffers reflection at the inner surface at point R to strike the outer surface at point T from where it continues to suffer multiple reflections. The length of the arc PT is equal to the skip distance k for the pipe and PRT is equal to the path length l. Considering the triangle PRS having sides PS = $D/2$, SR = $D/2 - t$, and RP = k, the sine law shows that:

$$1/2\sin\theta = (D/2 - t)/\sin\alpha = D/[2\sin\alpha + \theta], \qquad \text{(D.1)}$$

since

$$\sin(180° - \alpha - \theta) = \sin(\alpha + \theta),$$

from which it can be seen from the second and third expressions that:

$$\sin(\alpha + \theta) = (D \sin\alpha)/(D - 2t) = \sin\alpha/(1 - 2d), \qquad \text{(D.2)}$$

where $d = t/D$: the ratio of wall thickness to diameter. The angle θ (the angle subtended by one half of the path length to the centre) is thus given by:

$$\theta = \sin^{-1}[\sin\alpha/(1 - 2d)] - \alpha. \qquad \text{(D.3)}$$

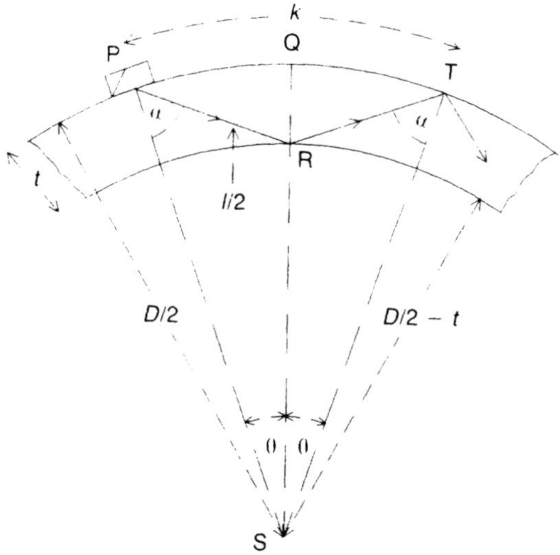

(a)

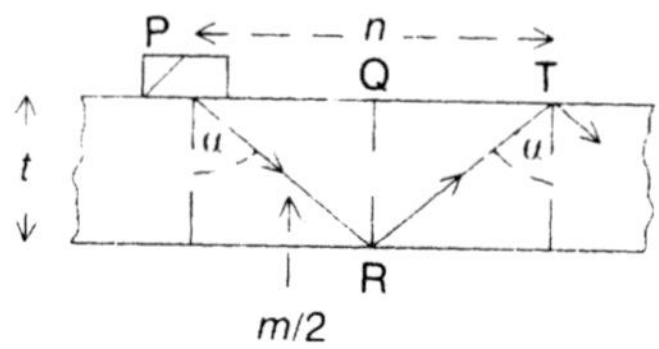

(b)

Fig. D.1 Sections of (a) pipe wall and (b) plate having equal thicknesses (c.f. Figs 7.12(a) and (b)).

The equality of the first and second terms of equation D.1 shows that the path length, l, is given by:

$$l = (D - 2t)\sin\theta/\sin\alpha. \tag{D.4}$$

It can easily be seen that the ratio of the arc length k (the skip distance) to the circumference, πD, is equal to that of the angle 2θ to 360°, so that:

$$k = \theta D\pi/180°. \tag{D.5}$$

Figure D.1(b) illustrates the location of a shear wave probe of angle α at a point P on the surface of a parallel-sided plate of thickness t. Here the skip distance is PT $= n$ and the path length is PRT $= m$. It is seen from the figure that:

$$\tan\alpha = PQ/QR = n/2t, \tag{D.6a}$$

and:

$$\cos\alpha = PQ/QR = 2t/m, \tag{D.6b}$$

$$\text{giving skip distance} = n = 2t\tan\alpha, \tag{D.7a}$$

and

$$\text{path length} = m = 2t/\cos\alpha. \tag{D.7b}$$

Thus the skip distance ratio is given by:

$$f_s = k/n = \theta\pi/(360°d\ \tan\alpha), \tag{D.8a}$$

and the path length ratio by:

$$f_p = l/m = (1 - 2d)\sin\theta/2d\ \tan\alpha, \tag{D.8b}$$

where θ is defined by equation D.3 and $d = t/D$.

Mode conversion caused by beam spread from a compression-wave probe located on the curved surface of a metal cylinder (section 7.4)

Figure 7.18b depicts a circular cross-section, centre O, of a cylinder tested by the pulse-echo method. A compression wave probe is located at point A on the curved surface and the beam angle is sufficiently wide for mode conversion to take place on lateral reflection at point B which has been selected in view of the reflected shear wave crossing the axis AE of the probe, i.e. the diameter of the circular cross section, at right angles at C. On reaching D, the reflected shear wave which is converted back to the compression mode which returns to the source at A. The wave pattern for this particular path is a symmetrical one.

Let α_L be the angle between the longitudinal ray AB and the radius OA. Because two of the sides of the figure ABO are of the same length (they are radii) the incident angle at B is also equal to α_L. The requirement for symmetry is that angle BCA should be 90°. Putting α_T as the angle of reflection for the mode converted shear ray BD, it is seen that ABC is a right-angled triangle in which the angle ABC is equal to $\alpha_L + \alpha_T$. Because the sum of the angles in a triangle is equal to 180°, we have:

$$2\alpha_L + \alpha_T = 90°,$$

$$\sin\alpha_T = \sin(90° - 2\alpha_L). \qquad \text{(E.1)}$$

Using the trigonometrical relation;

$$\sin(x - y) = \sin x \cos y - \cos x \sin y,$$

and putting $x = 90°$ and $y = 2\alpha_L$ and taking into account that $\cos 90° = 0$ and $\sin 90° = 1$, we have:

$$\sin(90° - 2\alpha_L) = \cos 2\alpha_L. \qquad \text{(E.2)}$$

From the trigonometrical relation:

$$\cos 2x = 1 - 2\sin^2 x$$

it is seen from the equations E.1 and E.2 that:

$$\sin \alpha_T = 1 - 2\sin^2 \alpha_L. \qquad \text{(E.3)}$$

From Snell's law (equation 2.48):

$$\sin \alpha_T / \sin \alpha_L = c_T / c_L, \qquad \text{(E.4)}$$

where c_T and c_L are the speeds of the respective shear and compression waves.

Substituting equation E.4 into equation E.3 gives the quadratic equation:

$$2\sin^2 \alpha_L + (c_T/c_L)\sin \alpha_L - 1 = 0$$

for which the solution is given by:

$$\sin \alpha_L = \tfrac{1}{4}\{- c_T/c_L \pm [(c_T/c_L)^2 + 8]^{1/2}\}. \qquad \text{(E.5)}$$

Because $\sin \alpha_L$ is positive for values of α_L ranging from 0° to 180°, the right hand side of equation E.5 must also be positive, so that:

$$\sin \alpha_L = \tfrac{1}{4}\{[(c_T/c_L)^2 + 8]^{1/2} - c_T/c_L\}. \qquad \text{(E.6)}$$

APPENDIX F

Transit time for ultrasonic waves in reinforced concrete (section 10.2.4)

With reference to Fig. 10.3 it is seen that:

$$\mathrm{TR} = l,\ \mathrm{TP} = \mathrm{QR} = a/\cos\theta \quad \text{and} \quad \mathrm{PQ} = 1 - 2a\tan\theta.$$

Because the angle of refraction at P is 90° for an angle of incidence θ, we have from Snell's law (equation 2.48) that $\sin\theta = c/c_s$ where c and c_s represent the ultrasonic speeds in the concrete and steel bar respectively. Hence:

$$\cos\theta = (1 - \sin^2\theta)^{1/2} = (c_s^2 - c^2)^{1/2}/c_s, \tag{F.1}$$

and

$$\tan\theta = \sin\theta/\cos\theta = c/(c_s^2 - c^2)^{1/2}, \tag{F. 2}$$

thus

$$\mathrm{TP} = \mathrm{QR} = a/\cos\theta = ac_s/(c_s^2 - c^2)^{1/2}, \tag{F.3}$$

and

$$\mathrm{PQ} = 1 - 2a\tan\theta = l - 2ac/(c_s^2 - c^2)^{1/2}. \tag{F.4}$$

The transit time, t, is then given by:

$$t = \mathrm{TP}/c + \mathrm{QR}/c + \mathrm{PQ}/c_s = 2a(c_s^2 - c^2)^{1/2}/cc_s + l/c_s. \tag{F.5}$$

Substituting $t = l/c$ in this equation gives:

$$c^2[4a^2 + (c_s t - l)^2] = 4a^2c_s^2,$$

from which it can be seen that:

$$a/l = [(c_s - c)/4(c_s + c)]^{1/2}. \tag{F.6}$$

Useful addresses

American Society of Nondestructive Testing (ASNT), 1711 Arlingate Lane, PO Box 28518, Columbus, OH 43228-0518, USA.

American Society for Testing and Materials (ASTM), 196 Race Street, Philadelphia, PA 19103, USA.

British Institute of Non-Destructive Testing, 1 Spencer Parade, Northampton NN1 5AA, UK.

British Standards Institution, 389 Chiswick High Road, London W4 4AL, UK.

Deutscher Industrie Norm (DIN), Burggrafenstrasse 4–7, Postfach 11707, D-1000, Berlin 30, Germany.

European Committee for Standardization (CEN), Central Secretariat, rue de Stassart 36, B-1050 Brussels, Belgium.

International Organisation for Standardisation (ISO), 1 rue de Varembe, Casse Postage 56, CH-1211 Geneva 20, Switzerland.

Personnel Certification in Non-destructive Testing, 1 Spencer Parade, Northampton NN1 5AA, UK.

The National Nondestructive Testing Centre, AEA Technology, 521 Harwell, Didcot, Oxfordshire OX11 0RA, UK.

TWI (formerly known as The Welding Institute), Abington Hall, Abington, Cambridge CB1 6AL, UK.

APPENDIX H

Notes on units

SI (Système International) units, which are used throughout this book, are based on the units of mass, length and time as being the metre (m), kilogram (kg) and second (s). Multiples and fractions of these units are related to unity by powers of 10^3 using the following prefixes:

pico (p) $= 10^{-12}$
nano (n) $= 10^{-9}$
micro (μ) $= 10^{-6}$
milli (m) $= 10^{-3}$
kilo (k) $= 10^{3}$
mega (M) $= 10^{6}$
giga (G) $= 10^{9}$
tetra (T) $= 10^{12}$

For example: 1 nm (nanometre) $= 10^{-9}$ m
1 μs (microsecond) $= 10^{-3}$ s
1 GHz (gigahertz) $= 10^{9}$ Hz

Units and symbols used in the text are given in Table H.1.

Table H.1 Units of quantities used in the text

Quantity	*Symbol*[a]	*Unit*
Mechanical and acoustical units		
Mass	m	kg (kilogram)
Length/displacement	l	m (metre)
Radius	r	m
Time	t	s (second)
Area	A or S	m^2
Density	ρ	$kg\,m^{-3}$
Speed/velocity	v or c	$m\,s^{-1}$
Attenuation coefficient (distance)	α	m^{-1}, neper m^{-1}, dB m^{-1}
Attenuation coefficient (time)	α'	s^{-1}, neper s^{-1}, dB s^{-1}
Frequency	f	Hz (hertz = s^{-1})
Angular frequency	ω	s^{-1}
Acceleration	a	$m\,s^{-2}$
Force	F	N (newton = $kg\,m\,s^{-2}$)
Torque/couple	T	N m
Work/energy	E	J (joule = N m)
Power	P	W (watt = $J\,s^{-1}$)
Pressure, stress	p or σ	Pa (pascal = $N\,m^{-2}$)
Elastic modulus	q	Pa
Compressibility	β	Pa^{-1}
Mechanical impedance	Z_m	$N\,m^{-1}\,s$
Stiffness	S	$N\,m^{-1}$
Compliance	C_m	$m\,N^{-1}$
Mechanical resistance	R_m	$N\,m^{-1}\,s$
Specific acoustic impedance	Z_a	rayl (=$kg\,m^{-2}\,s^{-1}$)
Characteristic impedance	ρc	rayl
Power (acoustic)	W	W (watt)
Intensity (acoustic)	I	$W\,m^{-2}$
Energy density	E	$W\,m^{-3}\,s$
'd' coefficient	d	$C\,N^{-1}$
Electrical and magnetic units		
Potential difference/EMF	V	V (volt)
Current	I	A (ampere)
Charge	Q	C (coulomb = A s)
Polarization	P	$C\,m^{-2}$
Impedance	Z	Ω (ohm)
Resistance	R	Ω
Inductance (self)	L	H (henry)
Inductance (mutual)	M	H
Capacitance	C	F (farad)

[a] The symbols used here are mainly those recommended by the International Electrotechnical Commission (IEC) and the British Standards Institution (BSI) but it is inevitable that the limited extent of the Roman alphabet may give rise to clashing. In this instance it is necessary to use alternative symbols which may be arbitrarily chosen.

Index

Lightning Source UK Ltd.
Milton Keynes UK
15 May 2010

154189UK00001B/83/A

9 780412 604706